Ethiopic Astronomy and Computus

Ethiopic Astronomy and Computus

OTTO NEUGEBAUER

THE RED SEA PRESS
TRENTON | LONDON | NEW DELHI | CAPE TOWN | NAIROBI | ADDIS ABABA | ASMARA | IBADAN

THE RED SEA PRESS
541 West Ingham Avenue | Suite B
Trenton, New Jersey 08638

The Red Sea Press Edition © 2016

This book is being republished by The Red Sea Press by the kind permission of the Austrian Academy of Sciences.

ISBNs: 978-1-56902-440-9 (PB)
978-1-56902-441-6 (HB)

PREFACE

To the Memory of
N. Rhodokanakis

In 1906 N. Rhodokanakis published, in Vol. 151 of the Sitzungsberichte der Akademie der Wissenschaften, a catalogue of the Ethiopic manuscripts in the collection of the Nationalbibliothek, Wien. On several occasions in this catalogue Rhodokanakis referred to astronomical and calendaric passages which suggested to me the possibility of a relationship with hellenistic methods that I was investigating at that time as part of a larger program concerned with the history of ancient mathematical astronomy. Hence it seemed reasonable to study texts of this type in greater detail, preferably in cooperation with an Ethiopic scholar. Having failed to obtain professional help, I began to familiarize myself as far as possible with the linguistic and lexicographical background and fairly soon had the satisfaction of reaching an insight into the structure of some "astronomical" treatises, in particular the shadow tables and the pattern of the "Gates", i. e., topics related to "arithmetical methods" familiar to students of ancient astronomy.

Continued search in the catalogues of manuscripts during the past fifteen years led to a large collection of treatises of greatly variable length and contents, almost all of them unpublished. These texts, copied and recopied over a period of many centuries, concern for the most part the "Computus", i. e., the determination of the dates of the movable Christian and Jewish feasts, in particular Easter and Passover. I was soon led to conclude that this material could not be arranged in a systematic fashion without making entirely arbitrary decisions as to what came first and what followed. It is for this reason that I have chosen an alphabetical arrangement by topics. This leads to unavoidable repetitions but makes it easier to place any text of calendaric or astronomical content in its proper surrounding. A detailed index (p. 256) and index of passages cited (p. 247) should help in following a specific topic in its various manifestations.

A much needed step would be the edition of some of these treatises. In order to facilitate such a task I have given ample (but by no means complete) references to passages in manuscripts, though I realize that it will take years before a reader can assemble the relevant collection (and many more texts that I have not seen).

My transcriptions will offend the eye of the specialist. In order to save labor and expense I have dispensed with all phonetic symbols, considering it to be the only purpose of a transcription to enable the user to restore the original writing. Hence I have omitted all long marks except for *ā* and *ē* and I have used freely the simple *e* instead of indicating the various phonetic qualities. I hope the reader will pardon these shortcomings in exchange for my efforts to make sense of texts which have been cruelly mistreated by chronological experts. That much remains to be done cannot be more evident to any reader than to myself.

Finally, I owe a special debt of gratitude to Miss Elizabeth Shushu Horton who completed the typing of the whole manuscript under the most adverse personal conditions.

Princeton, New Jersey
June 1978

OTTO NEUGEBAUER

TABLE OF CONTENTS

I.

INTRODUCTION

la scienza cronologica
è stata sempre privilegio
di paucissimi
P. Mauro da Leonessa,
Cappuccino

The material here investigated originates from two different classes of sources: the primitive astronomy of the "Book of Enoch" on the one hand, and the cleverly arranged "Computus" of the church calendar on the other. These two groups of sources are separated from each other not only chronologically but also geographically: the Book of Enoch was composed in the last centuries B. C. in the Palestinian area (also represented in the Qumrân archives) while the Computus reflects conditions centuries later within Alexandrian christianity.

The "Enoch" material is easy to describe. A linear scheme for the variable length of daylight with constant difference 1 and a ratio 2 : 1 of longest to shortest daylight (hence daylight + night = 18 parts); a division of the year into four equal seasons of 91 days each (hence 1 year = 364 days) beginning with spring; finally a schematic lunar calendar of alternating full and hollow months (hence a lunar year of 354 days), associated with a pattern for risings and settings, again based on a simple arithmetical scheme.

It could seem tempting to assume Babylonian influence in these arithmetical devices. In fact, however, there is nothing to support such a conjecture. Neither the Babylonian ratio 3 : 2 for the length of daylight nor any trace of a lunar, solar, or planetary theory is found in the Book of Enoch, and this at a time when Babylonian astronomy had reached its highest level.

The background for the Computus (ḥasāb) is far more difficult to analyze. Obvious is the Alexandrian milieu: the "year" is the Alexandrian year as instituted by Augustus (beginning with Thoth 1 = August 29/30) while the chronological eras are derived from the era Diocletian (A. D. 284). The 532-year cycle which underlies the whole calendaric system is the consequence of combining the 7-day week with the ("Callippic") 76-year cycle, i. e. with the Alexandrian calendar and with the 19-year luni-solar cycle. This latter cycle originated in Babylon but is also well known to Greek astronomers ("Metonic cycle"). In Babylonian astronomy the 19-year cycle was of great practical importance because of its connection with a real lunar calendar, i. e. a calendar that operated with true syzygies. In the Hellenistic-Roman calendar of Egypt the true lunations play no role whatsoever. Consequently the 19-year cycle in combination with julian years is not an element of Alexandrian computational astronomy, as we know it in great detail, e. g. from the "Handy Tables" of Ptolemy and Theon. It is ridiculous to praise the use of the 19-year cycle for the Easter computus as a survival of the scientific Alexandrian tradition. In fact it is nothing but the return to a simple luni-solar cycle, well known for many centuries. It seems plausible to consider the Alexandrian Jewish community as the carrier of this tradition. At any rate we know from the texts discussed here that the Jews made use of a schematic lunar calendar based on the 19-year cycle but adapted to the Alexandrian civil calendar.

Thus, the 532-year cycle is not a new scientific discovery but the trivial consequence of combining the Alexandrian calendar with the Jewish (and planetary) week and with the 19-year cycle: 532 is the smallest common multiple of 4, 7, and 19. All this is not a matter of astronomical investigation or insight but part of church politics which remains at the core of all Easter rules and controversies.

There is good evidence for the assumption that the Alexandrian Easter computus reached its final form in the early 5th century, based on the writings of Annianus (around 412) and Panodorus who adapted the commonly used era of Diocletian to the mystical 5500 years between the creation and the birth of Christ. Again religious speculations are here at work, not astronomy. Probably in the same time a new era was initiated, called era of "Grace", that conveniently took its departure from a moment 76 years after the beginning of the era Diocletian. A similar process took place in the 6th century also in the Latin church in adopting the 19-year cycle and transforming the era of Diocletian into the "Christian era".

If the astronomical background of the 532-year cycle is trivial, the practical arrangement of the tables which had to provide the user with the dates both of Passover and of Easter and of the associated feast days shows real skill and understanding of the arithmetical structure of the relevant numerical sequences. The basic concept is, of course, the 11-day interval of the "epact" (e) in relation to the 19-year cycle which necessitates the insertion of one 12-day interval (the "saltus lunae", placed at the end of the cycle). On the epact depends the date of the Jewish New Years Day, the maṭqeʿe (m), which determines the dates of all moveable Jewish feasts, in particular Passover ($p = m + 190$). The Christian feast, however, necessarily involves also the sequence of the days of the week, hence the cycle year c and the weekday of a fixed day in the year, here Thoth 1 = Maskaram 1, called "solar epact" or τῶν Θεῶν (in reference to the planetary week), i. e. ṭentyon (t). Depending on c and t one can tabulate the dates of Easter, or, more conveniently, the date of an earlier feast, called "Nineveh" (n), so that all other moveable feasts have a fixed distance from it, in particular Easter ($f = n + 69$). This is exactly the contents of the 20 columns of twenty-eight 19-year tables: the characteristics t of the Alexandrian years, the parameters e and m for the Jewish, and n for the Christian feasts, leading to p and f.

Modern scholars, knowing the boundaries of the Easter dates, postulated by the Alexandrians (repeatedly conflicting with the "Roman" wishes), had no difficulty in reconstructing the "Alexandrian" Easter table and confirming it from historically attested dates. This table has been published several times, e. g. in Ginzel, Hdb. III (p. 411 ff.). Unfortunately these modern tables express all dates in terms of the Roman julian calendar and thus completely obscure the underlying arithmetical structure. Reversion to the Alexandrian calendar reveals exact agreement with the Easter dates of the Ethiopic tables. Obviously these tables have preserved for us in all details the Easter tables of the Christian church of Alexandria in the fourth and fifth centuries.

By the time of the fourth century it had become a central feature of the Easter computus to keep the Easter Sunday securely separated from the Jewish Passover on the 14th of Nisan. Consequently it is not surprising also to find in the Ethiopic tables civil as well as lunar dates for Jewish feasts, in particular New Year and Passover. The inclusion of such dates makes sense, however, only when they represent actual facts with respect to the Jewish calendar practiced at the same period, i. e., in the first century of the era Diocletian. In other words the Ethiopic tables also reveal the structure of the Jewish calendar both in terms of the Alexandrian calendar and in terms of a schematic lunar calendar regulated by a 19-year cycle.

We have thus gained insight into an intermediary step in the remarkable history of this cycle. Discovered in the fifth century B. C. in Babylonia it was used to regulate a real lunar calendar in combination with a highly sophisticated theory of solar and lunar motion, capable of predicting the days of the first visibility of the new crescent. Greek astronomy of the hellenistic and Roman period in Alexandria had no interest in such problems since it operated exclusively with the simple and highly convenient schematic Egyptian calendar. In terms of Egyptian years the 19-year cycle had no advantage (a 25-year cycle takes its role) and the problem of lunar calendars had lost all interest. Only the Jewish community, as we now know, had adhered to its tradition by having its festivals determined by a 19-year cycle adjusted to the Alexandrian civil calendar, expressed in the "epact" computation. Since early christianity simply followed the Jewish Passover computus the 19-year cycle also entered automatically the Easter computus in spite of attempts, mainly in the Latin church, to operate with other cycles. But the adherence to lunar phases made such attempts doomed to failure and associated finally the 19-year cycle (since the 6th century) with the Western custom. The antagonism between Jews and Christians stimulated also among the learned rabbis (perhaps around the 6th or 7th century) the construction of a "reformed" Jewish calendar which attempted to determine the lunar phases independently of any civil calendar of the Gentiles. In the 12th century the "Code" of Maimonides approached this problem with all the theoretical tools provided by Arabic and Greek astronomy and with concepts closely reminiscent of Babylonian procedures. Thus the 19-year cycle finally returned to an astronomical level from which it had started more than a millennium earlier, while its most simplistic form remained embedded in the ḥasāb of the Ethiopians as well as in the "computus" of Byzantium and of the Latin Middle Ages.

NOTATIONS

1. Calendaric Parameters and Feasts

asc	Ascension (ʿergat or baʿāla 40)
bf	Beginning of Fast (ba'ata ṣom or derāra ṣom)
c	cycle number for years, from 1 to 19
ce	Inner of Earth (leba meder)
d	solar cycle (ʿawda ḍaḥay)
e	epact ('abaqtē)
ε	Evangelist (from 1 to 4: M., Mr., L., J.)
ep	epagomenal day (from 1 to 6)
f	Easter Sunday, date (fāsikā); $\bar{f}$: date in evening epoch (luna)
ho	Hosanna, Palmsunday, date (hosāʿena)
i	indictio
j	Day of John, weekday of I 1 from 1 (= Sunday) to 7; (ʿelata yoḥanes)
J.	(year of) John ($\varepsilon = 4$)
j.d.	julian day number
ju	Judith (yodit)
k	cycle number for years, from 1 to 532
kk	kēkros, 0;1
kn	kentros
L.	(year of) Luke ($\varepsilon = 3$)
le	ledata 'abaqtē
lu	luna, i. e. age of moon (*lu* 14 = full moon)
m	maṭqeʿe; $\bar{m} = 1$
M.	(year of) Matthew ($\varepsilon = 1$)
md	Prayer for Salvation (mehlā deḫnat)
mo	Mount Olive (dabra zayet)
Mr.	(year of) Mark ($\varepsilon = 2$)
mw	weekday of *m*
n	Nineveh (nanawē)
ν	lunar date of Maskaram 1
p	Passover, date; $\bar{p} = 14$
pt	Pentecost (baʿāla 50)
pu	Purim, Fast of Salvation (ṣoma deḫnat)
pw	Passover, weekday (ʿelata feśḥ)

ra	Synod (rakeb)
s	solar cycle, from 1 to 28
sh	Fast of Apostles (ṣoma ḥawāryāt)
t	ṭentyon, weekday of I 1 from 1 (= Wednesday) to 7
tb	Tabernacle (maṣalat), $\overline{tb} = 15$
tw	weekday of I 1 (in words)
w	weekday of any given date
yk	Yom Kippur (badr or ṣoma 'ayhud); $\overline{yk} = 10$
a, *b*, *c*	integers
[*a*/*b*]	integer part of the quotient *a*/*b* (example: [7/2] = 3)
$a \equiv b \bmod c$	the difference $a - b$ is a multiple of *c* (example: $65 \equiv 5 \bmod 30$; $21 \equiv 0 \bmod 7$)
‗‗‗	date in "evening epoch", i. e. lunar dates
┈┈┈	dates of Easter and Passover that fall into month VII

2. *Eras*[1]

A	julian years from Hijra
A. D., B. C.	julian years as used in modern historical chronology
AH	islamic years of Era Hijra
D	Diocletian or Martyrs (samāʿet)
G	Years of Grace (meḥrat)
J	Years of Incarnation (śegāwē), Ethiopic norm, not A. D.
W	Years of the World, or From Adam (ʿālam); Ethiopic norm
*	intercalary year

[1] For numerical data cf. s. v. Eras (p. 115).

II.

CATALOGUE OF TOPICS

Abbreviations of Technical Terms

In the column headings of tables one often finds abbreviations which simply consist of the initial syllables of the complete words (e. g. ʿāw qama for ʿāwda qamar[1]). Similar abbreviations occur in rules which repeat indentical phrases many times[2]. There exist, however, also abbreviations which are so short or which consist of internal or final syllables that their meaning can only become clear through the understanding of their function. Such abbreviations are nowhere listed in the dictionaries or in the literature and thus can become a major obstacle for the interpretation of a text or a table. It is for this reason that I give here a list of such non-trivial abbreviations which I encountered in my material.

ʾa	ʾatārd	Mercury (Arabic)
ʿā	ʿāmat	year
ʿā	saʿāt	hour
ʿā ḍa	ʿāwda ḍaḥay	solar cycle (28 years)
bo	ʾalbo	zero
da	daḫārāy	second (month)
da ḍa	ʿāwda ḍaḥay	solar cycle (28 years)
ḍu	ḥeḍuḍ	hollow month (29 days)
ʿe	ʿelatātihu	its days
fāzagā	fāsikā za magābit	Easter falls in month VII
gā	magābit	month VII
gā tā		cf. gā, cf. tā
ʾi	ʾiyobēl	jubilee (49 years)
lā qa	lāʿelāy qamar	latest date for Easter (i. e. VIII 30)
lafā	lafāsikā	for Easter
lafāzagā	lafāsikā za magābit	of Easter falling in month VII
lami	lamiyāzyā	of month VIII

---

1 Cf., e. g., Rhodokanakis [1906] p. 82; Vat. Catal. I p. 484; Hammerschmidt, Ṭānā I p. 157 f.

2 E. g. BM Add 16217 14ᵇ: baʿā(ta) ṣo(m) la(ʾema) ra(kabka).

lu	melu'e	full month (30 days)
māḫ	māḫfad	"tower" (zodiacal sign)
māḫ	māḫtam	"seal" (76-year period)
na	?	ordinary lunar year (12 months)
ña	nāfaqa (?)	½
parā	parāqlitos	Pentecost
san	sanbat	"week" (7·19 years)
sar, ser	sager	6th epagomenal day
qa	qadāmāy	first (month)
su	subā'ē	"week" (7 years)
tā		= tā qa
tā qa	tāḥtāy qamar	earliest date for Easter (i. e. VII 26)
ṭaf	ṭafe'e	invisible (new moon)
ṭe	ṭer	month V
zagā	zamagābit	for month VII

Abū Shāker, "Computus" [1]

The date of this work is well established from references in the text. The incipit mentions as the upper limit of the treatise[2] the year

W 6750 = Alex. 1569 = A. H. 655 (= A. D. 1257 /8)

and in ch. 15 the number of days since creation is computed[3] for

Saturday Yakātit (VII) 3 D 972 = Muḥarram 1 A. H. 654
(= A. D. 1256 Jan. 29).

In ch. 29 the year J 1200 is used as epoch[4].

Except for some small gaps, at least three copies of the "Computus" are extant:

BM Add 16252: Dillmann, BM Cat. No. 36 (p. 40—44). This is a copy made for Krapf[5] (around 1840)[6]

Abbadie, Cat. No. 140 (p. 155) = Conti Rossini [1915] No. 211 (p. 195 f.) = Chaine, Abbad. No. 140 (p. 87).

Vat. 106: Vat. Cat. pp. 386—391; p. 392 note.

Incomplete copies are:

BM 809 to 813 (Wright, Cat. Nos. 383 to 387).

Dillmann, Berlin HSS No. 83,3[7] (p. 73); in Amharic

Strelcyn, Cat. Griaule IV, No. 388,1 (p. 21); only a few chapters.

The "Computus" itself seems to be comparatively well preserved. A list of the headings of its 59 chapters is given in the Vatican Catalogue (Vat 106, pp. 386—391) and a slightly more condensed version is found in Dillmann's B. M. Catal. (No. 36, pp. 40—44). Neither one of these lists gives an adequate picture of the actual contents since they mainly reproduce the initial sentences of the chapters. The Computus is written, however, in a very prolix style and one can never know what is, and what is not, the contents of a chapter that begins with "if you wish to know . . ." Only a critical edition with a commentary could remedy this deplorable situation.

---

1 Cf. also below s. v. Authors (p. 22).

2 Dillmann, BM. Cat. p. 40/41 (from BM Add 16252); also Conti Rossini [1915] p. 195 (from Abbadie No. 140, p. 155 f.).

3 Dillmann, BM. Cat. p. 42. Cf. also below s. v. Eras, p. 125.

4 BM Add 16252, 139^b, 9.

5 Dillmann, BM. Cat. p. 40, p. 44.

6 Fumagalli BE p. 22, No. 172.

7 Or. oct. 238 (not 236 as given by Conti Rossini [1899, 2], p. 636).

In present Ethiopic usage the term "Abū Šāker" has become synonymous with "calendar"[8] and modern scholars more or less follow this habit by postulating Abū Shāker's authorship for whatever treatise concerns calendaric matters. This attitude is facilitated by the fact that neither the "Computus" nor the bulk of calendaric treatises scattered among our manuscripts have been published. My own investigation of these sources lends no support to the assumption of Abū Shāker's ultimate authorship.

Appendix. Summary of some Chapters of the "Computus"

Colophon[9].

Chapter 1[10]. Introduction. Different calendaric norms.

Chapter 8[11]. Diverse remarks on calendaric matters (Coptic, Roman, Greek, Syrian norms for months, epochs, etc.). Names of the months in the Almagest (without realizing the identity with the "Coptic" names) in part affected by misreading of the Arabic.

Chapter 9[12]. Five "foundations"; 1 and 2 concern the determination of the weekday at the end of a given interval, 3 to 5 dealing with three eras of given distances between their epochs.

Chapter 10[13]. From the Hakemī zīj of Ibn Yūnis (around A. D. 990) on the Deluge in relation to the Eras Nabonassar, Philip, Alexander, Augustus, Diocletian, Hijra, and Yezdegerd.

Chapter 11[14]. The year according to the Islamic calendar; remarks on sexagesimal fractions of the day.

Chapter 21[15]. Ending with intercalations and some intervals of 79 days in the Alexandrian and the Roman-Syrian calendars.

Chapter 22[16]. Omitted by the scribe because of duplicating ch. 21.

---

8 Cf., e. g., Ewald [1844] p. 184, n. 5. For this reason also references in manuscripts should not be taken at face value (e. g. Princ 5884 2[b] III, 5; 3[a] III, 17; etc.).

9 Published and translated in Dillmann, BM. Cat. p. 40/41. From BM Add 16252.

10 Partially translated in Dillmann, BM. Cat. p. 41.

11 BM Add 16252 25[a], 20 to 28[b], 13.

12 Ibid. 28[b], 14 to 31[a], 7.

13 Ibid. 31[a], 8 to 32[b], 12.

14 Ibid. 32[b], 12 to 33[a]; I have no complete copy.

15 Ibid. 115 to 118[b], 15; I have no complete copy.

16 Ibid. 118[b], 16 to 119[a], 2; cf. Dillmann, BM. Cat. p. 42.

Chapter 23[17]. Weekdays and 28-year "solar cycle". The difference of $1^1/_4$ days between Alexandrian and Enoch year is called "solar epact" (obviously in analogy to the lunar epact $365 - 354 = 11$ days—unrelated to the ṭentyon); thus $28 \cdot 1^1/_4 = 35^d$ as accumulation in 1 solar cycle. Arabic influences: "gates" = zodiacal signs, anomaly of solar motion, etc.

[17] Ibid. 119[a], 3 to 123[a], 5.

Alexander

See Eras (p. 123).

Amharic

Amharic plays a very secondary role in our investigations. Not only does this language appear late in Ethiopic literature[1] and then in small proportion to Geʿez[2], but when we find in our texts additions, or even short sections, in Amharic[3] they are usually coupled with Arabic astrology or astronomy, i. e. with material that is well-known and much better represented in other sources.

For one short Amharic treatise, written in 1896, and supposedly representing the "Guide for the Blind", see s. v. "Guide" (p. 165).

[1] Cf. e. g., Cerulli, Lett. p. 50/51; Doresse [1962] p. 145; Ullendorff, Ethiop. p. 155; etc.

[2] Macomber [1975] p. 401: "Only 23 of 500 manuscripts . . . have at least substantial sections written in Amharic". For a list of Amharic treatises, cf. Conti Rossini [1899, 2] p. 636 f. (in contrast to pp. 608—635 for Geʿez).

[3] For a glossary, cf. Littmann [1908].

Arabic

It is a well established fact that much of Ethiopic literature is an adaptation of Arabic versions, that originated in Palestine or Egypt. The amazing story of the transmission from French and German originals on the "Miracles of the Virgin" to Ethiopia[1] illustrates this process. And

[1] E. Cerulli, Il libro etiopico dei Miracoli di Maria e le sue fonti nelle letterature del Medio Evo Latino, Roma 1943. Also Cerulli, Lett. pp. 81 to 99.

the Arab conquest in the sixteenth century had a profound influence on the later history of the country and its relations to Europe.

A totally different situation prevails, however, for the astronomical and calendaric material in Ethiopic texts. These are completely dominated either by the Enoch tradition or by the Alexandrian "Computus". This does not exclude an occasional addition of Arabic material[2]—always recognizable by transliterations of Arabic terminology, e. g., of the names of the planets or the zodiacal signs and the lunar mansions[3]—but never assimilated to the traditional topics. Hence Enoch's twelve "gates" of the moon's risings and settings coexist with their Arabic misinterpretation as the twelve zodiacal signs[4] which contributes to the disorder that prevails in most treatises.

Some short sections (usually with Amharic terminology) display accurate parameters from Arabic lunar theory. I know of two, slightly different, sets which are based on the following values for the mean synodic month and for the "solar year":

A[5]: $1^m = 29;31,50,7,57,30^d$ B[6]: $1^m = 29;31,50,8,9,24^d$

$1^y = 365^d5;49^h = 365;14,32,30^d$ $1^y = 365;15^d$.

The length of the tropical year given in A is known from Ibn Yūnis (about A. D. 990)[7]; the value for the lunar month, however, is not known to me from any published source. The value in B is ascribed by Bīrūnī again to Ibn Yūnis "who follows the Chaldeans"[8]. The same number is found in a Coptic treatise of 1768[9].

[2] E. g. Arabic (i. e. "Hindu-Arabic") numerals, between Ethiopic numerals, in tables (Vat Borg 12 81^b).

[3] Planets: Vat 54 88^a—89^a; Vat 119 57^a I, 16—59^b II, 19. Zodiacal signs and lunar mansions: Vat 119 62^b I, 13—64^a II, 18; Vat 128, Strelcyn [1955] p. 88—91. And many other instances.

[4] This situation has been completely perverted by Charles who declared (among other absurdities) "the writer [of the Book of Enoch] is acquainted with the signs of the zodiac, but carefully refrains from mentioning them, replacing them by his system of portals" (Charles, Enoch, p. 190).

[5] In Vat Borg 1 3^b, 1—15.

[6] In Vindob 6 3^a, 1—9; Guidi [1896] 367 I, 17—19 (Marḥa 'Ewur); BM Add 16217 25^a I, 1—6 (including the tithi $1^\tau = 0;59,3,40,16,18,48^d$); as a later addition in EMML 215 4^a II, 6/7.

[7] Communication by Prof. Kennedy: this value is found in Kāshī's Khāqānī zīj and in the zīj of Shams al-Munajjim, who credits it to Ibn Yūnis. — Cf. also Copernicus, Commentariolus (trsl. Rosen p. 65/66) and Tycho Brahe (Opera II p. 15, 15).

[8] Communication by E. S. Kennedy.

[9] Sobhy [1942] p. 179 (Sect. 16); cf. s. v. Coptic p. 74.

For both groups exact arithmetical consequences are derived from these parameters, e. g. the values for the epact or for the length of the 19-year, 28-year, and 532-year cycles. The corresponding approximations are very good. For the 19-year cycle one finds from these data for the difference $\Delta = 235^{m} - 19^{y}$ the following values:

A: $\Delta = 0;5,3,40,12,30^{d} = 2;1,28,5^{h}$

B: $\Delta = 0;3,38,3,11^{d} \sim 1;27,13,20^{h}$.

The error could have been made exactly 0 in A by assuming a year of $365;14,48,28,57,30^{d} = 365^{d}\,5;55,23,45^{h}$. In B the error would be $0;0,0,0,0,1^{d}$ for a year of $365;14,48,31,24,41^{d}$.

None of all these parameters has the slightest influence upon the surrounding Ethiopic texts which are satisfied with explaining the epact of 11 days as the difference between 365^{d} and 354^{d}.

Armenia

Armenian calendaric usages are occasionally mentioned as well as Hebrew, Roman, and Syrian variants[1].

At the end of one of the "rearranged 19-year tables"[2] is written[3] zarome waza'armānyā "from Rome and from Armenia". The tables are of the ordinary type and surely of ultimate Alexandrian origin.

---

[1] EMML 215 49ᵃ II, 28; 49ᵇ I, 25.

[2] Cf. s. v. Calendar, p. 36.

[3] BM 816 38ᵃ, below the table.

Astrology

Astrology, in the technical sense of the word, requires the determination of planetary positions and data of spherical astronomy, hence computational techniques and auxiliary tables. No trace of such things has been found anywhere in Ethiopic texts.

Under Arabic influence, however, some elementary concepts of astrological doctrine make their appearance, remaining unassimilated fragments from a foreign culture. Only on the very margin of astrology lie traces of an early Graeco-christian tradition which associates the "Prophet Ezra" with astrological predictions and makes him the author of a "ḥasāb" for lucky and unlucky days[1]. Faint reflections of astrological lore are also visible in the Book of Jubilees[2] and in the Slavonic Enoch[3]. Abraham "observing the stars" all night long in order to discover "the character of the year with regard to the rains" has an Ethiopic parallel in a "Computus of the Rainy Season" (keramet) which is based on gematria: "You add the ṭentyon[4], the 5 epagomenal days, and the single extra days[5]; then reduce (the total) modulo 7[6]. If the remainder is 1 . . ." etc.

As examples for the intrusion of conventional hellenistic astrological concepts can be mentioned the pairing of signs known as "looking at each other"[7], or planetary rulers assigned to the four seasons of the year[8] and the consecutive hours of the day[9].

A short chapter on astrological geography (each one of the seven planets ruling certain countries, e. g., Saturn Ethiopia, Nubia, India, etc.) is found in one manuscript appended to calendaric tables; in another text written on the margin of a diagram for the lunar mansions[10]. Similarly, diagrams concerning the lunar phases carry marginal notes with completely foreign astrological contents (e. g., about the four temperaments or the four elements)[11]. Nevertheless such additions were faithfully copied from manuscript to manuscript.

In the "Book of Jubilees" one finds a few references to "signs (ta'amra) of Heavens" or "of sun, moon, and stars"[12]. The connection with the above-mentioned "observations" of Abraham suggests, however, weather prognostications, familiar from Greek works of the type of Theophrast's Περὶ σημείων (around 300 B. C.)[13]. Once also "omina

---

[1] For the Greek material, cf. Nau, ROC 12 (1907) pp. 14—21 and the notes to these texts in CCAG 8, 3 pp. 3, 26, 86, 88 and 8, 4 p. 80. For the Ethiopic "jours fastes et néfastes", cf. Grébaut, ROC 18 (1913) p. 97 f.

[2] VIII, 3 (Charles, p. 66 f. and note) and XII, 16 (Charles, p. 94).

[3] Cf. s. v. Enoch, Appendix, p. 113.

[4] Misinterpreted by Grébaut as "first day of the rain" (!).

[5] From 364 to 365 ('elata meryā); cf. Dillmann, Lex. 169. Also Upps 3 61[b], 5 to 13.

[6] tegadef baba 7-tu. Cf. also s. v. Gematria, p. 162.

[7] Vindob 6 32[b] II, 10—16 (as in Bouché-Leclerq, AG, p. 161, Fig. 17 I).

[8] Vat 54 88[a]—89[a].

[9] Vat 128 103[b] II; Vat 119 55[b] I, 10—19. Cf. also Vat. Catal. p. 485.

[10] Berol 84 20[b] and BM 816 6[b], respectively.

[11] BM 809 4[a] = BM 816 17[a].

[12] Jubilees IV, 17 and XII, 17 (Charles, p. 37, p. 94). All this has nothing to do with zodiacal signs (contrary to Charles' notes, p. 37, p. 55).

[13] Cf., e. g., Pauly-Wissowa, RE Suppl. VII, col. 1412—1415.

(sagala) of sun and moon and stars" are mentioned[14], probably also prognostications and not astrological in the strict sense.

Grossly misleading is the terminology used by Milik who, in discussing a hypothetical "Book of Noah", assumes in it "an astrological section giving a series of horoscopes"[15]. The only basis for this description is a fragment of a Hebrew (cryptographic) manuscript from Qumrān published by Allegro[16]. In it bodily characteristics of a person are described in a fashion which could perhaps belong to an astrological treatise though a "House of Light" and a "Pit of Darkness" do not fit such an origin. An "Ox" or "Bull" need not be the zodiacal sign Taurus[17]. Of course, there is no trace of one, much less of a "series" of horoscopes.

[14] Jubilees VIII, 3 (Charles, p. 67).

[15] Milik, Enoch, p. 55 f.

[16] John M. Allegro, Discoveries in the Judean Desert of Jordan. V. Qumrān Cave 4. Oxford 1968, pp. 88—91. Also Allegro [1964].

[17] Milik considers as terminus ante quem the end of the 4th/beginning of the 3rd centuries B. C. for this "Book of Noah" which would be very early in view of what we know about the history of astrology. Cf., e. g., Sachs [1952].

Authors

Texts concerning the determination of feast days and chronological matters, known as ḥasāb, i. e. Computus, range from a few paragraphs to relatively extensive treatises (e. g. Berol 84). Many of the shorter texts and calendaric tables are written on one or two pages left blank at the end of some manuscript. Even in the larger treatises very little logical order can be seen in the presentation of the material. It is my impression that this whole Computus literature represents fragments from introductions or commentaries to the 532-year tables, comparable to the more or less contemporary introductions to the "Handy Tables" by Theon of Alexandria (fifth century) and others[1]. Connections with a specific author, e. g., Demetrius (early third century), are extremely dubious.

An entirely different type of source is represented by the works of Arabic-Coptic church historians and theologians of the tenth to the thirteenth centuries, some of which were eventually translated into Ethiopic (e. g., Abū Shāker in the late sixteenth century). Among these authors are Saʿīd ibn Baṭrīq (around 930), al-Muqaffaʿ (960), Abū Shāker

[1] The existence of similar Latin treatises supports this conjecture; cf. below s. v. Calendar, Appendix II, p. 65.

(1250), Girgis ibn ʿAmīd (1275). Writers of that period derive their astronomical knowledge, whatever it may be, from Arabic sources. Their influence on Ethiopic calendaric methods and concepts is practically nil (at least as far as I can see), in spite of their fame as "authorities"[2].

Abu'l Barakāt

Priest at the church of al-Muʿallaqa in Old Cairo. He died about 1324/25. His theological writings deal also with historical and calendaric matters. Cf. Graf, CAL II pp. 438—444.

Abū Shāker

An-Nušū' abū Shāker ibn Buṭrus ar-Rāhib[3] ('abušaker walda 'abi 'elkaram)[4] was "diaconus ecclesiae S. Mariae in, Moallaca, in Cairo vetere"[5]. For his writings, theological and others, (belonging to the period from about 1250 to 1280)[6] cf. Graf, CAL II, pp. 428—435.

In Europe, since the seventeenth century, Abū Shāker was considered the author of the "Chronicon orientale", preserved in Arabic. In 1931, however, Chaine disproved this assumption[7]. Thus there remains for the present context only a Computus (ḥasāb) to be considered, a treatise translated from the Arabic into Ethiopic in the second half of the sixteenth century[8]. For this work the authorship of Abū Shāker seems fairly well established, although the Arabic original (with the title tārik, i. e. "history")[9] is lost[10], except, perhaps, for some anonymous fragments[11].

---

[2] I see no basis for Chaine's pronouncement (Chron. p. 10) that Egypt and Ethiopia "durant toute leur vie littéraire" followed the same authorities.

[3] Also called Petrus ibn Rāhib (e. g., Chaine, Chron. p. 11, n. 1).

[4] From the incipit of BM Add 16252, Dillmann, BM. Cat. p. 40.

[5] Dillmann, BM. Cat. p. 40/41; Weld Blundell, Roy. Chron. p. 494, n. 1.

[6] Cf. Graf, CAL II, p. 428. Conti Rossini [1899, 2] p. 206, associates him with the time of the Patriarch Cyrillos III (1235—1243), a period of "refiorimento letterario".

[7] Cf. also Graf, CAL II, p. 433. According to Chaine [1931] the Chronicon is based on an older historical work by Girgis ibn ʿAmīd (around 1270). Cf. also below s. v. Girgis p. 24.

[8] Graf, CAL II, p. 433.

[9] According to Abu'l Barakāt; cf. Graf, CAL II, p. 432.

[10] Cerulli [1973] p. 142.

[11] Chaine [1931] p. 392. A reference to the tārik (on the 70-year "weeks of Daniel") seems to fit ch. 20 of the Computus (Dillmann, BM. Cat. p. 42). I see no basis for Graf's connecting the History of Alexander with the Computus (CAL II, p. 433).

In one text[12] we find the following isolated remark; "Of Abū Shāker, the son of Yārēd, of the Province Gojjam[13], who emigrated from his country". Perhaps one should relate this note to a similar note about Girgis ibn ʿAmīd who is called "from Kebrān" (an island in Lake Ṭānā)[14].

Annianus

A monk, living in Alexandria around 400. He is credited (on the basis of statements by Syncellus[15a] around 800) with the expansion of the 19-year cycle to the 532-year cycle and the construction of a corresponding Easter table for the 11th cycle, i. e. for the years W 5853 to 6384, which is the first cycle of the era G (= J 353 to 884)[15b]. Ginzel's remark, "für die Ausbildung der Osterrechnung ist der Zyklus des Anianus nicht von Wichtigkeit geworden"[15c], was made without knowledge of the Ethiopic 532-year tables[15d].

Anonymous

Occasionally[16] one finds references to a "teacher"[17] as the authority for some computational procedure. No identification with some otherwise known author seems possible.

Athanasius

Bishop of Alexandria 326 to 373. In our context his Easter letters are of importance; cf. s. v. Easter, Appendix I (p. 98) and s. v. Ṭentyon (p. 219).

al-Baṭriq see Saʿīd al-Baṭrīq (p. 25)

Buṭrus ar-Rāhib see Abū Shāker (p. 22)

---

[12] BM Add 16217 8ª, I, below table (which is a 19-year table, the sixth in the sequence of seven).

[13] The area south of Lake Ṭānā.

[14] Cf. below p. 24.

[15a] Chronographia, ed. Dindorf, Bonn 1829, I, pp. 62—65; p. 597.

[15b] Cf. s. v. Eras, p. 117 Table I.

[15c] Ginzel, Hdb. III, p. 235.

[15d] Unger's pronouncement "Annianus war weiter nichts als ein handwerksmäßiger Passacalculator" (Pauly-Wissowa, RE 18, 3 col. 633, 35) has no meaning.

[16] BN 160 81ª I, 17; II, 18; 82ª I, 17; EMML 215 61ᵇ I, 11/12, I, 31, 62ª I, 30.

[17] Dillmann, Lex. 144: mamehr "doctor, magister, praeceptor".

Clement

A "qalēmenṭos" is quoted, together with Abū Shāker and Girgis ibn Amīd in Ṭānā 34[18].

Demetrius, Patriarch of Alexandria see Demetrius (p. 92)

Enoch see Enoch (p. 109).

Eutychius see Sa'id ibn Baṭrīq

Girgis (Georgios) ibn 'Amīd[19]

He was born 1203/5 in Egypt[20] and died in 1273 in Damascus. His World History (maṣhafa tārik) was written after 1260[21]. In the manuscript Ṭānā 53[22] he is called giyorgis walda 'āmid zakebrān, i. e. from the island Kebrān in Lake Ṭānā. A translation to Ethiopic of his "History" was completed on Maskaram 1 of the year D 1011 ($\approx$ A. D. 1294 August 29)[23]. His work is probably the basis of the "Chronicon orientale" that was incorrectly ascribed to Abū Shāker[24].

Many references to Girgis are found in our texts[25]. In some of the calendaric tables one finds among the marginal notes the remark, "here in the year of the Martyrs 1035, Georgis and Abūšaker finished their book"[26]. Which book we are not told. The date in question (A. D. 1318/19) is so late[27] that it can at best concern the completion of the translation to Ethiopic.

---

18 Fol. 29^b ($k = 100$) and 37^a ($k = 930$). Cf. also Dillmann, BM. Cat. No. 36, p. 44, note m.

19 "The elder". For al-Makīn (i. e. the monk) Girgis ibn al-'Amīd ("der Jüngere") who wrote at the end of the 14th century; cf. Graf, CAL II p. 297.

20 Graf, CAL II p. 348: 1205 or 1203; Weld Blundell, Roy Chron, p. 494: 1223.

21 Graf, CAL II, p. 348/9.

22 Ṭānā 53 4^a top; cf. Hammerschmidt, Ṭānā I, p. 207.

23 Cf. Conti Rossini [1915] p. 194 f. ($\approx$ Abbadie No. 68).

24 Chaine [1931]; Graf, CAL II, p. 433. Cf. above s. v. Abū Shāker (p. 22).

25 E. g. BM 815 17^a I, 5 $\approx$ BM 827 120^a I, 5; Ṭānā 34, 29^b—92^b.

26 From BM Add 19658 54^b, I, at $k = 433$. A parallel is found in BN Abb 123, 195^a at $k = 435$ but it gives 1033 as the year. Neither D 1033 nor 1035 correspond to a cycle year 433 or 435. A plausible emendation would be 1043 (misreadings of 30 for 40 are not rare) which would correspond to G_2 435 = J 1319 $\approx$ A. D. 1326/7.

27 Another entry in BN Abb 123, 193^b at $k = 366$ gives for the "beginning of Abušāker's writing" the date G_2 366 = J 1250 $\approx$ A. D. 1257/8 in good agreement with his lifetime.

John, Bishop of Nikiu[28]

Author of a chronicle, written shortly before 700[29], extant only in an Ethiopic version, though originally written in Greek with some chapters in Coptic[30]. The Ethiopic translation was made in A. D. 1601[31] from an Arabic version. None of the original versions seems to have survived.

The Ethiopic text was edited and translated to French by Zotenberg [1883]. For an English translation, cf. Charles, Chronicle (1916).

Al-Muqaffaʿ

Severus ibn al-Muqaffaʿ[32], Bishop of Ashmunein[33]. Author of a history of the Alexandrian Patriarchs, written around 955.

Panodorus see s. v. Eras (p. 119)

Petrus ibn Rāhib see above s. v. Abū Shāker (p. 22)

Philip see s. v. Greeks (p. 165)

Saʿīd ibn Baṭrīq

also called Eutychius, born 877; Melkite Patriarch of Alexandria, 933—940[34]. Repeatedly quoted by Abū Shāker[35].

Severus see Al-Muqaffaʿ (above)

Synesius cf. s. v. Greeks (p. 165)

---

[28] In the Delta, NW of Cairo, tentatively identified with Ebšādi by Berg, Karte (IV, D6). Ptolemy's coordinates given in Geography IV 5, 49 (Nobbe, p. 259) in relation to Naukratis lead to a location on the western branch of the Nile at $\varphi = 30;45$. Cf. also Kees in Pauly-Wissowa, RE 17, 1 col. 342—344.

[29] Cf. Graf, CAL I, p. 470 f.

[30] Zotenberg [1883] p. 126.

[31] Cf. the colophon, Zotenberg [1883] p. 342/587 (maltreated by Charles, Chronicle, p. 202). From this colophon we obtain the following data for the completion of the translation: Ṭeqemt (II) 22 W 7094 (text: 7594!) = J 1594 = D 1318 (= A. D. 1601 [not 1602!] Oct. 19), Monday. For the Hijra date cf. s. v. Era p. 125. Some astronomical elements also given in this colophon are rather inaccurate.

[32] CSCO 52; cf. also Graf, CAL II p. 295, p. 300.

[33] Ancient Hermopolis (near Antinoopolis; $\varphi \approx 27;45$). Cf. also Graf, CAL II pp. 300—317; Chaine, Chron. p. 11.

[34] Cf. Graf CAL II pp. 32—38; Chaine, Chron. p. 27, p. 252 (Index p. 332).

[35] E. g., in ch. 23 and in ch. 30 (BM Add 16252 120[b], 9 f.; 141[a], 11; 19/20 etc.).

"Benediction" (barakat)

See Magic.

Bizan

Famous monastery (Debra Bizan) near Asmara, founded in A. D. 1390/91[1], repeatedly mentioned, e. g., in the "Liber Axumae"[2]. For an "Era Bizan" cf. s. v. Era (p. 123). A "Computus of Bizan"[3] is concerned with the chronology of Christ and other chronological matters particularly at the time of Susenyos[4].

---

1 Assfalg-Krüger, WB p. 392.

2 CSCO 58, p. 26, 10; 51, 5; 86, 13 etc.

3 BN Abb 37 1 ᵇ lower margin 5.

4 Year 23 = A. D. 1629.

Books

(See also s. v. Authors; Computus; Dates of Treatises.)

The term "book" (maṣḥafa) applies to a great variety of treatises. The largest works, e. g. the Book of Enoch, the Book of Jubilees, or the Didascalia, contain not more than passing references to astronomical or calendaric matters, excepting the chapters 72 to 82 in the Book of Enoch[1]. To what extent titles like "Ocean of Computus" (bāḥra ḥasāb)[2] or "Book of Computus" (maṣḥafa ḥasāb)[3] represent independent chronological works is difficult to say[4]. A work called zēnā 'aresta ḥasāb ("Report, or History, on the Introduction to the Computus" ??) is extant in two versions[5]; no author is mentioned.

---

1 These chapters probably belong originally to an independent treatise.

2 Wright, Br. Mus. Cat. p. 291 No. 384 fol. 2b; also Chaine, Chron. p. 26/27 note.

3 Berol 84 colophon 1 ᵃ, 3 seems to imply that the title refers to the whole manuscript, which, however, is by no means a uniform treatise.

4 The history of such works can be very complex; cf. e. g., below s. v. "Guide for the Blind" (p. 165).

5 EMML 215 2ᵃ ff.; Vindob 6 23ᵃ ff.

Calendar

Table of Contents

Explanatory Note

Two closely related topics, "Calendar" and "Feasts", lie at the center of interest for our discussion. The division of material between these two headings is obviously arbitrary to a certain extent but as a guiding principle may be considered the association of "calendaric" matters with the 532-years table and its arithmetical structure, whereas the "feasts" concern the moveable feasts, both Jewish and Christian, and their interrelations in terms of the Jewish lunar calendar and the Alexandrian "civil" calendar.

For the sake of greater flexibility I have separated "Eras" and "Cycles" from the two main articles. Similarly special sections are reserved for the key parameters connected with the Christian and with the Jewish New Years Day ("Ṭentyon and Day of John" and "Maṭqeʿe and Epact," respectively).

The calendaric procedures in the Jewish and Christian communities at Alexandria are not only the basis for the Ethiopic calendar but they are also important for the understanding of the Easter controversies between Alexandria and Rome. For this reason I have included some "Latin" material, e. g., in the article "Easter" the tables of Dionysius Exiguus (6th century) in relation to the time of Athanasius (4th century), or "concurrentes" and "regulares" in connection with the ṭentyon[1], or the "octaeteris" in contrast to the 19-year cycle[2].

[1] Cf. s. v. Calendar, App. II (p. 64—66); s. v. Ṭentyon p. 220.
[2] Cf. s. v. Cycles (p. 83—85).

1. General Structure

All basic elements of the Ethiopic calendar were well known long before the conversion of Ethiopia to Christianity. The calendar itself is a replica of the Alexandrian calendar, instituted by Augustus as a "julian" modification of the Egyptian ("moveable") calendar. The seven-day week is of Judeo-Christian origin (unrelated to the Hellenistic planetary week, at least as far as terminology is concerned). The 19-year cycle is deeply embedded in the latest phase of Babylonian astronomy and therefore also in the subsequent Hellenistic astronomy ("Metonic cycle"). Finally the fundamental eras of Ethiopic chronology are based on (or identical with) the Era Diocletian, including its epoch date Diocletian 1, Thoth 1[3]. Hence there is not a single new astronomical concept in the structure of the Ethiopic calendar. Its origin lies clearly in Roman Egypt.

A significant feature of the Ethiopic calendar, as we have it, is the strict separation between Jewish and Christian feasts, both classes being completely determined as soon as one single parameter is known: the maṭqeʿe (m), the date of the Jewish New Years Day, and the date of Nineveh (n) for the Christian feasts. With m is coordinated the date of Passover ($p = m + 190$), and similarly with n the date of Easter Sunday ($f = n + 69$). Consequently the whole calendar is determined as soon as the dates for m and n are given.

The theological and historical arguments which produced the final rules for the relative positions of festivals do not concern us here. It is of importance, however, to emphasize the conclusion that the data for Jewish festivals and for the underlying schematic lunar calendar must correspond to actual contemporary usage, i. e., Jewish usage at the time when this whole schema was designed in Alexandria. Since it is the avowed aim in the Easter Computus to avoid any coincidence of the Jewish Passover with Easter Sunday, it would have been absurd to construct a pattern for the "Jewish" feasts not coinciding with the actual Jewish practice. Hence our calendar treatises not only provide us with the Christian calendar but inform us at the same time about the structure of the Alexandrian Jewish calendar (which, of course, by no means must be identical with Jewish practice in the Palestinian-Syrian area, or other communities in the Diaspora).

---

[3] Friday, August 29, A. D. 284. The other eras ("World", "Incarnation", "Grace") are defined in relation to the Era Diocletian with calendarically convenient differences (cf. s. v. Eras, p. 116).

2. Calendaric Tables

It follows from the foregoing remarks that all our calendaric data can be found as soon as three fundamental parameters are known: the maṭqeʿe (m) with its 19-year period of the lunar phases, the ṭentyon (t) for the week-day of Maskaram (I) 1, and the "Evangelist" (ε) with the period of 4 years for the "julian" intercalation pattern of the Alexandrian civil calendar. Tables based on these parameters must have as their period the lowest common multiple of 19, 7, and 4, i. e., their product 532.

Many copies of such tables are extant, giving for 532 consecutive years a list of interrelated calendaric parameters. One can describe these tables as a periodic matrix of 532 lines, all elements of which are known as soon as the first line is known, which, in turn, is determined by the values of the three parameters m, t, and ε (or equivalent parameters, e. g., the epact e instead of m). As mentioned before these parameters are based on the Alexandrian calendar and the situation at the epoch date Thoth 1 of the Era Diocletian or one of the related eras (e. g. "World").

Our texts give many equivalent rules for finding the fundamental parameters for a given year, expressed in terms of any one of the four Ethiopic eras. In modern terminology these rules amount to the following:

The maṭqeʿe (m) is easiest found from the corresponding "epact" e on the basis of

(1) $$m + e = 30$$

where e is given by

(2) $$e = 11 \cdot r \bmod 30 \qquad \text{where} \qquad r \equiv \mathrm{W} - 1 \bmod 19$$ [4].

The ṭentyon (t) is determined by

(3) $$t \equiv \mathrm{W} - 1 + [\mathrm{W}/4] \equiv \mathrm{J} + [\mathrm{J}/4] \equiv \mathrm{G} + 2 + [\mathrm{G}/4] \bmod 7$$ [5].

In order to change the norm of t (according to which 1 = Wednesday) to the norm 1 = Sunday, the "Day of John" (j) is introduced by

(4) $$j \equiv t + 3 \bmod 7.$$

Finally the intercalation pattern is expressed in the naming of each quadruple of years by the Evangelists (ε), found from any one of the Ethiopic eras by

(5) $$\varepsilon \equiv \mathrm{W} \equiv \mathrm{J} \equiv \mathrm{D} \equiv \mathrm{G} \bmod 4$$

The residues $\varepsilon = 1, 2, 3, 4$ are associated, in this order, with Matthew,

[4] Cf. s. v. Maṭqeʿe and Epact, p. 190 (5).

[5] Here G can be replaced by D because G ≡ D mod 7.

Mark, Luke, John. The years of John ($\varepsilon = 4$) are the leap years with 366 days.

All remaining parameters can be derived from these few basic numbers[6]. The subsequent tables show their values for the first years of the four Ethiopic eras, and for the first four years of the eras W and G[7].

Table I

Year 1 of Era	W	c	ε	e	m	t	j	n	julian August 29
W	1	1	1	0	30	7	3	4	−5492
J	5501	10	1	9	21	1	4	26	8
D	5777	1	1	0	30	3	6	8	284
G	5853	1	1	0	30	7	3	4	360

Table II

W or G	c	ε	e	m	t	j	n
1	1	1	0	30	7	3	4
2	2	2	11	19	1	4	26
3*	3	3	22	8	2	5	16
4	4	4	3	27	4	7	30
5	5	1	14	16	5	1	22

The numbers *c* and *e* in Table I clearly demonstrate the exceptional character of the era J, being the only one not adjusted to the 19-year cycle. On the other hand it is obvious that W and G are arithmetically equivalent eras since they are exactly eleven 532-year cycles apart.

A. The 532-year Tables

The most common type of these large tables gives for each year of a complete 532-year cycle the values of some 20 calendaric elements,

[6] Cf. below p. 42 ff.

[7] D and G represent here any one of the 532-year cycles into which these eras are divided.

arranged in 28 sub-tables, each of which covers one 19-year cycle[8]. Usually year 1 corresponds to the first year of the third cycle of the "Years of Grace":

$$G_3\ 1 = W\ 6917 = D\ 1141 = J\ 1417\ (\approx \text{A. D. } 1424/5)^9.$$

Table III

	k	c	le	ep	tw	ε	j	t	e	m	mw	$\bar{m}$	yk	$\overline{yk}$	tb	$\overline{tb}$	n	bf	p	pw	$\bar{p}$	f
1	1	1	none	5	Tue.	Math.	3	7	none	30	4	1	9	10	14	15	4	12	10	5	14	13
	2	2	25	5	Wed.	Mar.	4	1	11	19	1	1	28	10	3	15	26	10	29	2	14	5
	3	3	14	5	Thu.	Luc.	5	2	22	8	7	1	17	10	22	15	16	30	18	1	14	25
	4	4	3	6	Sat.	Joh.	7	4	3	27	5	1	6	10	11	15	30	14	7	6	14	9
	5	5	22	5	Sun.	Math.	1	5	14	16	2	1	25	10	30	15	22	6	26	3	14	1
	6	6	11	5	Mon.	Mar.	2	6	25	5	1	1	14	10	19	15	12	26	15	2	14	21
	7	7	30	5	Tue.	Luc.	3	7	6	24	5	1	3	10	8	15	27	11	4	6	14	6
	8	8	19	6	Thu.	Joh.	5	2	17	13	5	1	22	10	27	15	16	30	23	6	14	25
	9	9	8	5	Fri.	Math.	6	3	28	2	2	1	11	10	16	15	5	22	12	3	14	17
	10	10	27	5	Sat.	Mar.	7	4	9	21	6	1	30	10	5	15	23	7	1	7	14	2
	11	11	16	5	Sun.	Luc.	1	5	20	10	5	1	19	10	24	15	13	27	20	6	14	22
	12	12	5	6	Tue.	Joh.	3	7	1	29	3	1	8	10	13	15	4	18	9	4	14	13
	13	13	24	5	Wed.	Math.	4	1	12	18	7	1	27	10	2	15	26	10	28	1	14	5
	14	14	13	5	Thu.	Mar.	5	2	23	7	6	1	16	10	21	15	9	23	17	7	14	18
	15	15	2	5	Fri.	Luc.	6	3	4	26	3	1	5	10	10	15	1	15	6	4	14	10
	16	16	21	6	Sun.	Joh.	1	5	15	15	1	1	24	10	29	15	22	6	25	2	14	1
	17	17	10	5	Mon.	Math.	2	6	26	4	7	1	13	10	18	15	12	26	14	1	14	21
	18	18	29	5	Tue.	Mar.	3	7	7	23	4	1	2	10	7	15	27	11	3	5	14	6
	19	19	18	5	Wed.	Luc.	4	1	18	12	3	1	21	10	26	15	17	1	22	4	14	26
2	20	1	none	6	Fri.	Joh	6	3	none	30	7	1	9	10	14	15	8	22	10	1	14	17
	21	2	25	5	Sat.	Math.	7	4	11	19	4	1	28	10	3	15	23	7	29	5	14	2
	22	3	14	5	Sun.	Mar.	1	5	22	8	3	1	17	10	22	15	13	27	18	4	14	22

Our Table III shows the arrangement of such tables; cf. also the photograph on Pl. I. Appendix I (p. 59—63) gives the data for one complete cycle in the era G or W. In our manuscripts *tw* and *ε* are usually written (in words) above each other in the same column. Similarly one finds the weekdays *mw* and *pw*, written in red, above *m* and *p* but not all tables give these weekdays and other small variants are common.

---

[8] This misled Dillmann to translate 28 'aqmar as "tabulae sunt 28", instead of "28 cycles" (BM. Catal. p. 44, No. 37, 1a).

[9] E. g., Princ 5884 1a—5b I; Ṭānā 34 3a—16b; Vat 119 71a—84b; Vindob 6 4a—17b.

In this table the numbers *k* and *c* count the years, *k* from 1 to 532, *c* always from 1 to 19. The Alexandrian calendar is represented by the number of the epagomenal days (*ep*) and the name of the Evangelist (ε). The weekday of the first day of the year is given three times: in *tw*, in *j* (Sunday = 1), and in *t* (Wednesday = 1). The most important parameters are the epact *e* and the related date *m*; for *le* cf. below p. 191. We then have the civil dates for the Jewish festivals Yom Kippur (*yk*), Tabernacle (*tb*), Passover (*p*), and their (constant) lunar dates ($\overline{m}$, $\overline{yk}$, $\overline{tb}$, $\overline{p}$) and the weekdays of *m* and *p* (*mw* and *pw*). Finally we come to the Christian feast-dates Nineveh (*n*), Beginning of Fast (*bf*), and Easter (*f*), the final goal of the Computus.

Many tables have a heading tārik, i. e. "historical events", above the left hand margin of the tables, containing notes on biblical and historical chronology associated with the respective cycle years, e. g., of accessions or deaths of kings and patriarchs[10]. This column, as well as the right margin, carries also a variety of tabular elements, e. g., the consecutive numbers from 1 to 76 of 7-year "weeks" (subāʿe), of 133-year "weeks" (sanbat), counted in the order of the Evangelists, 49-year "jubilees" (from 1 to 10), etc. In the last column, or in the right margin, for the Easter dates (*f*) one often finds a note gā and numbers from 1 to 48 for those dates *f* which belong to the month Magābit (VII). Similarly, the earliest and the latest dates permitted for Easter, VII 26 and VIII 30, are marked by tā qa and la qa, respectively[11]. Scribal errors, of course, abound in these marginal notes.

Smaller 532-year tables[12] also exist which list only the parameters

tw and ε *e* and *mw* *bf* *p* *pw* *f*

but surprisingly give no cycle numbers, neither *k* nor *c*. Comparison with the larger tables shows, however, that the first year of the smaller tables corresponds (mod 532) to the year G 153 = J 505, in which a 19-year cycle of the era G begins simultaneously with a solar cycle in the era J[13].

A slightly different arrangement of the 532-year tables is preserved in tables[14] which cover only the three 19-year cycles from Diocletian[15]

[10] E. g., BM 816; Vindob 6; Ṭānā 34 which remarks at *k* = 457 "impious Diocletian" (cf. s. v. Eras p. 117, Table I).

[11] Cf. s. v. Abbreviations, p. 14; also below p. 46.

[12] BM Add 16217 2^a—6^b I; BM Add 16220 1^a—5^b I; Princ 5884 1^a—5^b I; Ṭānā 53 110^a—[117^a] (cf. also 86^a I, 17—100^b).

[13] Cf. below s. v. Cycles (Solar Cycle) p. 88.

[14] Jerus Arm 3483, 193^b—197^a.

[15] Called "years of Grace" in the text, a frequently occurring incorrect terminology.

27 to 84, i. e., A. D. 310/11 to 367/8. The date of these tables cannot be changed by multiples of 532 years because they contain a column for the "indictio" (*i*) which raises the tabular period to 15 · 532 years. As shown below (s. v. Easter, App. I, p. 98), the same data appear in the Easter tables of Athanasius (which begin at D 44, i. e. A. D. 328) and thus prove the existence in the early fourth century of the 532-year tables in Alexandria.

Beyond the addition of a column for *i* we find in these tables also a column for the lunar date of the Beginning of Fast ($\overline{bf}$) and of Easter ($\bar{f}$), the latter being the "luna" of Easter Sunday[16]. In the columns for *m*, *yk*, and *tb* it is noted when these dates belong to the earlier of the two possible months (I or II), and similarly for *bf*, *p*, and *f*.

In contrast to the common 532-year tables these early tables omit the columns *le*, *tw* and *ε*, which are of little practical use. Indeed, *tw* is the equivalent of *ȷ* and the association of the years with the Evangelists is perhaps an Ethiopic invention. It is surprising, however, that no column for Nineveh (*n*) is given, though this parameter became the point of reference for all Christian feasts[17].

We have also rather useless excerpts from the 532-year tables, giving only, besides the cycle numbers, the weekday (*tw*) of Maskaram 1 (in words), the Evangelist of the year (*ε*), and the epact (*e*). Since the latter runs through the same sequence of values in each 19-year cycle, these tables are effectively only tables for *ε* and *tw*, two parameters without much interest and useless for calendar computations without other parameters. Nevertheless we have such tables with well over 7000 entries[18], which serve as the chronological skeleton for notes on biblical and secular history.

B. Enlarged 532-year Tables

We have copies of 532-year tables[19] which include a larger number of moveable festivals, both Jewish and Christian, than the ordinary

---

16 Cf. the excerpt from these tables s. v. Easter, App. I, p. 100.

17 Cf. s. v. Feasts, p. 132.

18 Ṭānā 34 29^{a}—92^{b} (Hammerschmidt, p. 160 f.). These tables cover 382 complete 19-year cycles, i. e., 13 532-year cycles and 4 76-year cycles. They break off at the year W 7265 (= J 1765 = A. D. 1772/3).

19 Berol 994 37^{b}—65^{a}; BM 816 1^{a}—5^{a}. The first text is complete with 28 tables, each one occupying two facing pages. Of the second text only the second half of table 24 (on fol. 1^{a}) and the tables 25 to 28 are preserved. These tables also carry a numbering from 6 to 9, so that table "1" is the twentieth table of the whole cycle (beginning with *k* = 362).

tables of about 20 columns. These "enlarged" tables tabulate 36 parameters in some 28 to 30 columns, several of which display two related parameters (e. g., civil dates and lunar dates). As usual the tables are preceded by a column with historical notes (tārik). In the following list of columns[20] the addition of "a" and "b" indicates that both data are listed in the same column, usually "b" written in red above "a".

2	k	7	e	13	bf	20	ra
3a	c	8a	m	14	mo	21	asc
3b	le	8b	mw	15	ho	22	pt
4a	ep	9	$\overline{m}$ = 1	16	p	23	md
4b	i	10a	yk	17a	pw	24	sh
5a	tw	10b	$\overline{yk}$ = 10	17b	$\overline{p}$ = 14	25	pu
5b	e	11a	tb	18	f	26	$\overline{pu}$ = 27
6a	t	11b	$\overline{tb}$ = 15	19a	ce	27	ju
6b	j	12	n	19b	$\overline{ce}$ = 8	28	$\overline{ju}$ = 4

Note the inclusion of a column for the indictio (*i*) among the chronological parameters; also that the order of the 17 feasts listed follows their chronological sequence:

*m**	*ho*	*pt*
*yk**	*p**	*md*
*tb**	*f*	*sh*
n	*ce**	*pu**
bf	*ra*	*ju**
mo	*asc*	

where * indicates a Jewish feast, hence associated with a fixed lunar date (śarqa lēlit). It is of interest to note that the feasts *ce**, *pu**, and *ju** are here included while they are rarely mentioned in calendaric treatises[21].

The parameters in these tables correspond to the era G. Fortunately also the indictio (*i*) is tabulated which determines the 532-year cycle uniquely[22] as the cycle G_3, i. e. as the cycle from J 1417 to 1949 (i. e. A. D. 1424/5 to 1956/7)[23].

---

20 The column $\overline{ce}$ = 8 is missing in Berol 994.

21 Cf. s. v. Feasts, p. 144—146.

22 Cf. s. v. Eras, p. 120 or p. 122, Table IV.

23 Hence *k* = 362, the beginning of the tables in BM 816, corresponds to A. D. 1785/6.

C. *The 19-year Tables*

This class of tables[24] lists only such parameters as depend on the weekday of Maskaram 1 (i. e., on t) and on the 19-year cycle (i. e., on c). Hence one needs 7 tables for $t = 1$ (Wednesday) to $t = 7$ of 19 lines each. Usually nine parameters are listed:

$$(1) \qquad c \quad t \quad j \quad e \quad n \quad bf \quad p \quad pw \quad f$$

in that order[25] (cf. Table IV).

Table IV

Wednesday

c	t	j	e	n	bf	p	pw	f
1	1	4	0	3	17	10	6	12
2	1	4	11	26	10	29	2	5
3	1	4	22	10	24	18	7	19
4	1	4	3	3	17	7	3	12
5	1	4	14	19	3	26	6	28
6	1	4	25	10	24	15	4	19

etc. -

c	t	j	e	n	bf	p	pw	f
17	1	4	26	10	24	14	3	19
18	1	4	7	26	10	3	6	5
19	1	4	18	17	1	22	4	26

Thursday

c	t	j	e	n	bf	p	pw	f
1	2	5	0	2	16	10	7	11
2	2	5	11	25	9	29	3	4
3	2	5	22	16	30	18	1	25
4	2	5	3	2	16	7	4	11

etc.

etc.

Tuesday

c	t	j	e	n	bf	p	pw	f
1	7	3	0	4	18	10	5	13
2	7	3	11	27	11	29	1	6
3	7	3	22	11	25	18	6	20
4	7	3	3	4	18	7	2	13

etc.

c	t	j	e	n	bf	p	pw	f
17	7	3	26	11	25	14	2	20
18	7	3	7	27	11	3	5	6
19	7	3	18	18	2	22	3	27

The numbers in these tables can be easily shown to agree with the numbers in the 532-year tables. The ṭentyon t is, of course, constant in each table, corresponding to the weekday given in words at the heading of each table; $j \equiv t + 3 \pmod 7$ is called "Beginning of Thoth (śarqa-tito)[26], i. e., weekday of Thoth = Maskaram 1 (in the norm Sunday = 1). The epact e and the date p of Passover depend only on the 19-year cycle, i. e. on c, whereas the weekdays pw of p depend also on t[27]. Finally, $bf = n + 14$, $f = n + 9$ mod 30, depending, therefore, like n, only on c and t. The columns for n are identical with Table XIII derived below p. 50.

[24] I made use of the following MSS: Berol 84 13^b—15^a I; Berol 991 103^b—106^b; BM 815 119^a—120^b I; BM Add 16220 7^a—8^b; BM Add 19658 60^a—61^b I; Princ 5884 9^a—10^b I; Vindob 6 19^b—21^a I. Incomplete copies in Berol 994 $76^{a,b}$; BM 754 4^a; BM 815 119^a—120^b I. The tables in Princ 5884 and Vindob 6 mention the "Greeks" (ṣereʿe) in the heading.

[25] Column n is omitted in Berol 84, Berol 991, BM 815. Instead, Berol 991 also gives mw and m.

[26] Incorrectly interpreted by Dillmann as "novilunium" (BM. Catal. p. 44 No. 37 1b). For tito = Thoth cf. e. g., the lists of month names in different languages, e. g. Princ 5884 12^a III, 17/18.

[27] Cf. for the reason p. 52 below.

Since 7 tables of one 19-year cycle provide only 133 sets of values of the sequence (1), each set must occur 4 times in the 532-year tables[28]. This is indeed the case as can be seen by direct comparison with the 532-year tables[29]. In this way one finds, e. g., for the table with $t = 1$ the following years k within one 532-year cycle:

Table V

c	k			
1	58	153	248	495
2	2	97	192	439
3	41	136	383	478
4	80	327	422	517
5	24	271	366	461
6	215	310	405	500

c	k			
17	131	226	321	416
18	75	170	265	360
19	19	114	209	304

$t = 1$

The numbers in each quadruple increase by 95 (mod 532) whereas numbers in consecutive lines can be associated in such a fashion that they decrease by 56 from c to $c+1$ (e. g. 248, 192, 136, etc.). The first mentioned difference of 95 years is known as the "cycle of the great epact" and both differences are clearely visible in the complete table for k as function of c and t (cf. below s. v. Ṭentyon p. 223 and Table II).

D. Rearranged 19-year Tables

We also have tables[30] which are arranged differently. These are, nevertheless, the equivalent of the 19-year tables described in the preceding section. These new tables arrange the values of 7 parameters in a small square, 7 of which are combined in a row corresponding to the weekdays from Wednesday ($t = 1$) to Tuesday ($t = 7$); 19 such rows are numbered in the left margin for the cycle years 1 to 19, as shown in Table VI. Thus the number of entries for each parameter is again $7 \cdot 19 = 133$.

[28] Therefore called manfaqa manfaqu la ʿābiyu qamar, i. e., "half of one half of the great cycle" (e. g., BM 816 38ᵃ below the table).

[29] Cf. below p. 59—63.

[30] Berol 84 19ᵇ—20ᵇ (upper part); BM 754 3ᵇ; BM 815 1ᵇ/2ᵃ; BM Add 16220 6ᵇ/7ᵃ; BM Add 19658 58ᵇ/59ᵃ; Bodl 36 146ᵇ; Princ 5884 8ᵇ/8ᵃ (in this order); Vindob 6 18ᵇ/19ᵃ.

Table VI

	Wedn.	Thurs.	Fri.	Sat.	Sun.
1	11 29 1 5 2 13 9	11 29 2 4 3 13 8	11 29 3 3 4 13 7	11 29 4 2 5 13 6	11 29 5 1 13
2	22 18 1 19 7 24 23	22 18 2 25 1 24 29	22 18 3 24 2 24 28	22 18 4 23 3 24 27	22
3	3 7 1 12 3 5 16	3 7 2 11 5	3 7	3	
	14				

etc.

(Sun.)	(Mo.)	(Tue.)	
		10	
3	18 22 6 28 2 20 2	18 22 7 27 3 20 1	(18)
10 15 3 2 19	30 10 6 14 4 2 18	30 10 7 13 5 2 17	(19)

Several texts add a twentieth row to Table VI, called "second computus" (kāleʿe ḥasāb), which should repeat the first one in order to indicate the periodicity of the table[31]. In the texts known to me this row is particularly marred by scribal errors. Obviously the absurd design increased the bewilderment of the scribes.

The parameters in each square represent

$$
\begin{array}{ccc}
e & & p \\
t & f & pw \\
\bar{v} & & bf
\end{array} \qquad (1)
$$

Most tables[32] give in the right margin a column headed "epagomenal (days)", displaying the sequence 5 5 5 6, etc. This makes no sense since the epagomenal days do not have a 19-year period.

The interpretation (1) of the numbers in the squares is not only based on their numerical values but it is confirmed by notes written below the tables or scribbled between the numbers. The simplest way of revealing the meaning of the numbers consists in restoring them to 7

[31] Omitted, e. g., in BM Add 16220 and in Bodl 36.

[32] Exception: Berol 84.

parallel columns for each weekday. Then one obtains, e. g. for the section shown in Table VI, a table of the ordinary type (cf. Table VII). The only new parameter is $\bar{v} = e + 2$, the lunar date of Maskaram 1[33]. Comparison with the 19-year tables (cf. above p. 35 Table IV) reveals two deviations: first, bf is given the value $f + 4$ instead of $f + 5$, as it should be[34]; secondly, the numbers associated in Table VII with $c = 1, 2, \ldots, 19$ actually belong to $c = 2, 3, \ldots, 19, 1$[35]. One could explain this discrepancy by assuming that we are dealing here with completed years whereas the cycle numbers c and k in the 532-year tables always refer to the current year.

Table VII

Wednesd.

c	t	e	$\bar{v}$	bf	p	pw	f
1	1	11	13	9	29	2	5
2	1	22	24	23	18	7	19
3	1	3	5	16	7	3	12

etc.

Thursd.

c	t	e	$\bar{v}$	bf	p	pw	f
1	2	11	13	8	29	3	4
2	2	22	24	29	18	1	25
3	2	3	5		7		11

etc.

etc.

etc.

c	t	e	$\bar{v}$	bf	p	pw	f
18	6	18	20	2	22	2	28
19	6	30	2	18	10	4	14

(Mond.)

etc.

c	t	e	$\bar{v}$	bf	p	pw	f
18	7	18	20	1	22	3	27
19	7	30	2	17	10	5	13

(Tuesd.)

However, the explanation of $bf = f + 4$ as an error is supported by a modified type of the present tables which use the correct relation $bf = f + 5$[36]. Similarly we have tables[37] which do not count the rows from 1 to 19 but, e. g., by a column for the epact which then begins correctly with 11, 22, 3, ... corresponding to $c = 2, 3, 4, \ldots$ Starting with $c = 2$ could be motivated by the desire to locate the "saltus lunae" not between the last row of a table and the first, but to put it ahead of

---

[33] Cf. s. v. Maṭqeʿe and Epact, p. 190.

[34] Cf. below p. 42.

[35] This is easiest to confirm for the columns e and p by comparison with the table "XIX" (below, App. I, p. 59).

[36] Cf. below p. 41.

[37] Cf. below p. 41.

the end, so that there could be no doubt about the periodicity of the whole pattern.

The whole set of the $7 \cdot 19 = 133$ squares of Table VI occurs four times in each 532-year table. The cycle years k which belong to a given square located in row c (corrected!) and in column t can be found directly below s. v. Ṭentyon, Table II, p. 224.

Modifications

Beyond the group of almost identical tables of the type shown in Table VI, p. 37, we have several isolated tables which in some way or another modify the list of parameters enumerated in (1) p. 37. Also, these modified tables begin their rows with parameters that belong to $c = 2$ and their columns with $t = 1$, i. e., Wednesday.

1. One variant[38] consists in expanding (1) to

$$(2) \qquad \begin{array}{ccc} e & & p \\ t & f & pw \\ \bar{p} & n & bf \end{array}$$

as shown for the left upper corner in Table VIII.

Table VIII

	Wedn.	Thursd.	Fri.
1	11 29 1 5 2 13 26 9	11 29 2 4 3 13 25 8	11 3 3 13 24
2	22 18 1 19 7 24 10 23	22 18 2 25 1 24 16	22
3			

Note that also here the cycle numbers should be 2, 3, etc. and $bf = f + 5 = n + 14$ instead of $f + 4 = n + 13$.

2. In another variant[39] the number of parameters in each square is reduced to 5. The left upper corner is shown in Table IX.

---

[38] EMML 2077 5ª/4ᵇ (in this order!).

[39] Vat Borg 12 81ᵇ.

Table IX

	Wedn.	Thursd.	Fri.	
29 VII	5 9 2 25 40	4 8 3 24 41	3 4 23	11
18	19 23 7 9 26	25 29 1 15 20	24 14	22
7	12 16 3 2 33	11 15 4 1		3

The first column on the left gives the dates p of Passover (for $c = 2, 3, \ldots$). On the right margin a corresponding column gives values for the epact. The central number in each square (incidentally, written with "Hindu-Arabic" numerals) represents the weekday of Passover (Sunday $= 1$), the left upper corner is the date f of Easter, as comparison with Table VI (p. 37) shows. The right upper number is always $f + 4$, which agrees with the same error for bf in Table VIII. Finally, the left lower corner is $f + 20$ instead of $n = f + 21$[40]. Thus we have the following pattern:

$$\begin{array}{ccc} f & & bf \\ & pw & \\ n & & ? \end{array} \tag{3}$$

The numbers in the right lower corner (r) vary exactly in the same pattern as the dates pw, so that in each line $r - pw =$ const. The values of r vary between 15 and 49, the units being presumably days. The same situation can also be described as $n + r = 36$ or 66 since the pattern for n[41] locates its discontinuities at the same places as pw, but only for a decreasing sequence, while r increases with pw.

3. It is not surprising to find square tables reconverted to an ordinary tabular arrangement. This was indeed attempted[42], but instead of using single lines for each c the scribe used a single column for each weekday (cf. Table X). In the process a new error was introduced by

---

[40] Hence we have correctly $bf \equiv n + 14 \mod 30$.

[41] Cf. below p. 49.

[42] In BM 754 $2^b/3^a$, combined with the ordinary arrangement in squares on fol. 3^b.

calling each of the second numbers "Day of John" (*j*) instead of ṭentyon (*t*). But the table from which he constructed his new pattern had the correct values for $bf = f + 5$.

Table X

c	Wedn.	Thrs.	Fri.
1	11 e	11 e	11 e
	1 j	2 j	3 j
	13 ȳ	13 ȳ	
	10 bf	9 bf	
	29 p	29 p	etc.
	2 pw	3 pw	
	5 f	4 f	
2	22 e	22 e	
	1 j		

etc.

4. An essentially new modification is found in a table for 8 parameters, assembled in rectangles of 2 lines[43]. The left upper corner of such a table is shown in Table XI:

Table XI

e	Wedn.	Thursd.	Fri.
11	26 10 5 29	25 9 4 28	24 8
	7 28 14 24	6 27 13 23	5
22	10 24 19 13	16 30 29 19	15
	21 12 28 8	27 18 4 14	
3	3 17 12 6	2	
	14 5		

[43] From Vat Borg 17, 1ᵇ—2ᵇ.

All parameters agree with the cycle numbers $c = 2, 3, \ldots$ (which are not listed in this table). The parameters in each rectangle are easily identifiable as

$$(4) \qquad \begin{array}{cccc} n & bf & f & ra \\ mo & ho & asc & pt \end{array}$$

All Jewish data (e. g., p or pw) are now eliminated and the dates of five Christian feasts are introduced which are normally not listed in the 532-year tables[44]. In contrast to the previously described tables all relations are correct (excepting of course, isolated scribal errors). Thus we have (mod 30)

$$(5) \qquad \begin{array}{cccc} bf = n + 14 & f = n + 9 & ra = n + 3 & \\ mo = n + 11 & ho = n + 2 & asc = n + 18 & pt = n - 2 \end{array}$$

as it should be.

3. *Calendaric Parameters*

bf. Beginning of Fast

The relevant term, commonly used in treatises and tables, is ba'ata ṣom, "Beginning of Fast", but also derāra ṣom, "Evening of Fast" can be found[45].

In relation to the date of Easter Sunday (f) one has

$$(1) \qquad bf + 55 = f \qquad \text{thus} \qquad bf \equiv f + 5 \qquad \text{mod } 30.$$

The connection with Niniveh (n) is given by

$$(2) \qquad bf = n + 14.$$

These three parameters, n, bf, and f, are of principal importance for the whole system of Christian feasts[46].

The dates of bf are restricted to the interval

$$(3) \qquad \text{VI } 1 \leqq bf \leqq \text{VII } 5.$$

The earliest permissible date, VI 1, is said[47] to occur four times in each

[44] Cf., however, the enlarged tables (above p. 34).

[45] E. g., Vindob 6 $18^{b}/19^{a}$.

[46] In the Eastern Church the interval from bf to f is called "Great Lent", but the 55 days of the Coptic Church are reduced to 48 days. Cf. Saliba [1970] p. 183, 14.

[47] E. g., Princ 5884 6^{a} III, 19 — 6^{b} I, 9.

532-year cycle. In the Era Diocletian these years are the years $k = 35$, 130, 225, 320[48]; in the era of Grace $k = 491$, 54, 149, and 244[49]. Indeed, these years are, as they should be, also the years of the earliest possible Easter dates (VII 26)[50]. In all cases one has $c = 16$.

Some tables[51] contain also the lunar dates $\overline{bf}$ for the Beginning of Fast. The number of days $bf - m$ between bf and m reaches over five months, two of which must be hollow. Therefore $bf - m = \overline{bf} - \overline{m} - 2$; since $\overline{m} = 1$, one has

$$(4) \qquad \overline{bf} = bf - m + 3.$$

c. *The 19-year Cycle*

In the tables the years of this cycle are listed under the headings ʿawda ʾabaqtē, "Epact Cycle"[52], or ʾudata warḫ, "Cycle of the Moon (or Month)".

In fact we have here a schematic imitation of the famous Babylonian luni-solar cycle which contains $19 \cdot 12 + 7 = 235$ synodic months and thus restores the phases of the moon on which in principle the Jewish calendar—and hence also the Easter computus—rests. The numbers c of the cycle, running from 1 to 19, are obtainable from the numbers k in the 532-year cycle as residues modulo 19.

Data which depend not only on c but also on the Alexandrian calendar (e. g. the "Evangelist" of a year) repeat themselves in a cycle of $4 \cdot 19 = 76$ years. This cycle is honored with a special name, "seal" (māḫtam), of unknown origin. Seven seals make one lunar cycle (532 years)[52a].

---

[48] Enumerated in Princ 5884 but without naming the era. Similar in EMML 215 59ᵇ II, 2—10.

[49] Note the constant difference of 95 years (mod 532) between these four dates. For the reason, cf. s. v. Ṭentyon, Appendix (p. 223).

[50] Cf. also below, s. v. Feasts, p. 149.

[51] Jerus Arm 3483, 193ᵇ—197ᵃ, column 14, next to column 13 for bf. Cf. below s. v. Easter, Appendix I, p. 100.

[52] Vat 119, 71ᵃ—84ᵇ; Vindob 6, 4ᵃ—17ᵇ. Example from the colophon Vindob 6, 18ᵃ = "lunar cycle 426, epact cycle 8". Indeed, $k = 426 \equiv 8 = c$ mod 19.

[52a] Neither Dillmann, nor Guidi or Grébaut list this technical meaning though it occurs frequently in our texts (e. g., BM Add 24995 27ᵇ I, 16; BN Abb 37 1ᵃ II, 31—35; Marḥa ʿE 365 I, 21; Princ 5884 7ᵃ II, 8—14; Ṭānā 53 107ᵃ III, 29—31). Mauro da Leonessa says "non so per quale ragione" (Diss. Cron. p. 91). Perhaps the completion of the cycle by seven "seals" suggested a parallel to Solomon's seals.

e. *Epact*

See below s. v. Maṭqeʿe and Epact (p. 187). The epact for a given year W can be found from

$$c \equiv W \mod 19 \qquad e \equiv (c-1) \cdot 11 \mod 30.$$

ε. *Evangelists*

Each year of the Ethiopic calendar is associated with the name of one of the four Evangelists, repeating cyclically the canonical sequence from the Gospels:

(1) Matthew Mark Luke John.

This sequence is called ʿāwda ʿāmat, "cycle of years"[53].

Since the intervals between the epoch years of the four Ethiopic eras[54] are all multiples of 4, the rule

(2a) $$\varepsilon \equiv E \mod 4$$

gives the same result for either one of these eras (E). Then one defines

(2b)

$\varepsilon = 0$ or 4:	John
1:	Matthew
2:	Mark
3:	Luke.

The years of John are associated with 6 epagomenal days[55]; cf. below s. v. Epagomenal Days (p. 113).

ep. *Epagomenal Days*

Cf. s. v. Epagomenal Days (p. 113).

f. *Easter*

The term fāsikā means the date of Easter Sunday (*f*), in contrast to feśḥ, the date of Passover (*p*).

In the Ethiopic, i. e. Alexandrian, norm Easter Sunday is restricted to an interval of 35 days (as are all moveable Christian feasts), bounded by

(1) $$\text{VII } 26 \leq f \leq \text{VIII } 30.$$

These boundaries are known as "lowest moon" (tāḥtāy qamar) and as

53 E. g., in the 532-year tables of Ṭānā 34, Vat 119, Vindob 6.

54 Cf. s. v. Eras p. 117, Table I.

55 E. g., BN Abb 37 2ᵃ I, 25—32. One also speaks about "intercalary days of John", ʿelat sagra yoḥanes, e. g. Upps 3 61ᵇ, 10.

"largest moon" (lā'elāy qamar)[56]. The lower limit thus keeps Easter at a date later than VII 25, the day of the vernal equinox, a fact which was eventually incorporated into the definition of Christian Easter[57].

For checking the tables, the following relations are useful:

(2) $$f = n + 69 \equiv n + 9 = bf - 5 \qquad \text{mod } 30$$

which connect the dates of the most important Christian festivals n, bf, and f.

Computation of f.

The date of Easter (f) in a given year is very easy to determine if one makes use of the rule that f is the next Sunday following Passover (p)[58]. Indeed, from the number of the year in any era, e. g. W, one can find the parameters c or e and t for the year in question[59]. Hence one also knows the date of the Jewish New Years Day

(3) $$m = 30 - e \qquad (\text{I } 15 \leqq m \leqq \text{II } 13)$$

and therefore the date of p because

(4) $$p = m + 190 \qquad (\text{VII } 25 \leqq p \leqq \text{VIII } 23).$$

But from t, the weekday of I 1, we also know the weekday mw of m and thus also the weekday

(5) $$pw = mw + 1$$

of p, which in turn determines the date f of the next Sunday.

Example[60]. Find f for the year G 470:

$$470 \equiv 14 = c \quad (\text{mod } 19) \text{ thus } 13 \cdot 11 = 143 \equiv 23 = e \quad (\text{mod } 30).$$

Furthermore t (cf. below s. v. Ṭentyon, p. 221) from

$$470 + [470/4] - 1 = 470 + 117 - 1 = 586 \equiv 5 = t \text{ (mod 7) i. e. Sunday.}$$

Consequently

$$m = 30 - 23 = (\text{II})\ 7 = \text{I } 1 + 36 \text{ and Sunday} + 36 = \text{Monday} = mw.$$

[56] E. g. in the 532-year tables BM 815 3a—16b. Also abbreviated to tā qa and lā qa, respectively. These terms were mistranslated by Zotenberg (B.N. Cat. p. 261, l and v) as "petit et grand cycle lunaire". — The earliest dates occur at $k = 491, 54, 149, 244$, the latest at $k = 27, 122, 217, 312$.

[57] Cf., e. g., CSCO 221/222 p. 115, 1/2 / p. 100, 32. Cf. also below s. v. Feasts p. 136.

[58] Cf. s. v. Feasts p. 133.

[59] Cf. s. v. Maṭqe'e and Epact p. 190 (5a) and s. v. Ṭentyon p. 221 (2a).

[60] Cf. also below p. 49 in the computation of n.

Furthermore

$p = \text{II}\,7 + 190 = \text{VIII}\,17 \quad 197 \equiv 1 \pmod 7 \quad \text{Monday} + 1 = \text{Tuesday} = pw.$

Hence the next Sunday is Tuesday + 5 and thus the date of Easter Sunday

$$f = \text{VIII } 17 + 5 = \text{VIII } 22.$$

All these data can be found in the 532-year tables (below p. 63 and Table "XIX", p. 59).

It should be noted that the tables usually give only the day numbers for f without naming the month. Since f may vary between VII 26 and VIII 30, the numbers from 26 to 30 are ambiguous. For this reason the manuscripts usually add the sign gā (for magābit = VII) to those dates which belong to the earlier month. But also without such a note the month of f is easily recognizable from the date of p which is unambiguously determined by (4) and thus leaves no doubt about the date of the next following Sunday. By the same procedure the apparent ambiguities in the dates of all movable Christian feasts can be eliminated.

$\bar{f}$. *Lunar Date ("luna") of Easter*

Since the lunar date $\bar{p}$ of Passover is always 14, one has for $\bar{f}$

(1) $$\bar{f} = 14 + (f - p)$$

and therefore also the limits

(2) $$15 \leqq \bar{f} \leqq 21.$$

Occasionally[61] texts refer to these boundaries as Magābit (VII) 15 and Miyāzyā (VIII) 21 respectively, a terminology which mixes the civil dates of the Easter limits with the lunar dates (which could at best only be referred to the lunar month Nisan).

The age $\bar{f}$ of the moon on Easter Sunday is known as "luna" to the Latin Computists. Also some of the Ethiopic tables, among them two which concern the first years of the Era Diocletian[62], tabulate the values $\bar{f}$ in a separate column. For the "luna" in Latin tables, e. g. the tables of Dionysius, cf. Krusch [1937] pp. 70—74 or p. 27 ff.

The boundaries given in (2) are characteristic for the Alexandrian Easter computus, in contrast to the Roman norm[63]

(3) $$16 \leqq \bar{f} \leqq 22.$$

---

[61] EMML 2077 4^b (which should follow 5^a).

[62] Jerus Arm 3483 193^b—197^a; Vat 1 209^b—210^b. A text with a column $\bar{f}$ based on the era G is BN Abb 123 184^a—197^b.

[63] Cf. Schwartz [1906] Ges. Schr. 5, p. 25/26 or Krusch [1937] p. 5, p. 12.

The difference between the rules (2) and (3) lies at the center of the "Easter Controversy".

As convenient checking rules can be mentioned

$$(4) \qquad \overline{bf} = \bar{f} + 4$$

and

$$(5) \qquad \bar{f} + pw = 22$$

which can easily be derived from the commonly used relations between bf and f, m and p, and the corresponding lunar dates.

j. *Day of John*

Weekday of the first of Maskaram, therefore also called śarqatito, "Beginning of Thoth"[64]. In contrast to the ṭentyon (t), which counts the days of the week from Wednesday = 1, the "Day of John" (j) is based on the norm Sunday = 1. Consequently

$$(1) \qquad j \equiv t + 3 \bmod 7$$

Period: 28 years. Mediaeval equivalent: "feria".

k. *532-year Cycle*

This cycle is commonly called "lunar cycle" (ʿāwda qamar)[65]. Also the term "remainder of the cycle" (tarafa qamar) is attested[66], which simply means "remainder modulo 532". Example: "in the year 7130 of Grace (actually W), and in the era Bizan 6674, and in the remainder of the cycle 214, and in the remainder of (the era) Bizan 290". Indeed $7130 \equiv 214$ and $6674 \equiv 290 \bmod 532$[67].

All calendaric parameters are the same for the same number k, regardless to which 532-year cycle of the era G they belong. If one operates with the era D the corresponding number k will be 76 higher than for G.

---

[64] Berol 84, 13ᵇ—15ᵃ; BN Abb 37, 2ᵃ I, 33 f.; Princ 5884 9ᵃ—10ᵇ; Vindob 6, 19ᵇ—21ᵃ. Cf. also below s. v. Ṭentyon and Day of John p. 219.

[65] E. g. in the headings of the 532-year tables.

[66] Wright, B. M. Cat. p. 221, from BM 754 182ᵃ; also CSCO 61 p. 26, 26. Guidi's translation "l'excédant (du comput) de la lune" (CSCO 66, p. 27, 2/3) is meaningless.

[67] From Wright, B. M. Cat. (cf. the prec. note). From $k = 214$ (thus $c = 5$) it follows $e = 14$, $m = 16$, $t = 7$ (not 6) as given in the text. For the era Bizan cf. below p. 123.

le. *Beginning of Epact*

The date called ledata 'abaqtē, "Beginning of Epact", can be obtained by the relation

(1) $$le \equiv m + 6 \bmod 30.$$

Period: 19 years. Cf. s. v. Maṭqeʿe and Epact (p. 191) for a tentative explanation of this parameter.

m. *Maṭqeʿe*

For a year W the civil date m of the Jewish New Years Day can be found from

(1) $$c \equiv \mathrm{W} \bmod 19 \qquad m \equiv (c-1)\cdot 19 \bmod 30.$$

Relation to the "epact":

(2) $$e + m = 30.$$

Period: 19 years. Cf. s. v. Maṭqeʿe and Epact (p. 187). The lunar date of m is by definition $\overline{m} = 1$.

mw. *Weekday of m*

The weekday of I 1 is known from column j (norm: Sunday = 1). The distance of m from I 1 is $m - 1$, if m belongs to month I, but $m + 29$ when m falls into month II. Since m is restricted to the interval

(1) $$\mathrm{I}\ 15 \leqq m \leqq \mathrm{II}\ 13$$

we have for the weekdays mw of m (in the norm Sunday = 1):

(2) $$mw \equiv \begin{cases} j + m - 1 \\ j + m + 1 \end{cases} \bmod 7 \begin{cases} \text{if } 15 \leqq m \leqq 30 \\ \text{if } \ 1 \leqq m \leqq 13 \end{cases}$$

Cf. also below s. v. pw (p. 52).

n. *Nineveh*

The date (n) of the feast of "Nineveh" (nanawē) is frequently used as the basic parameter for the determination of the dates of the other moveable Christian festivals. The reason for this exceptional role is perhaps the fact that n is the earliest moveable feast in the calendar year since it belongs to the interval

(1) $$\mathrm{V}\ 17 \leqq n \leqq \mathrm{VI}\ 21.$$

The relation with Easter is expressed by

(2) $$f = n + 69$$

from which it follows, since $69 \equiv -1 \bmod 7$, that Nineveh always falls on a Monday. Another useful relation is

$$n \equiv bf - 14 \bmod 30. \tag{3}$$

Computation of n

Exactly like the date f of Easter the date of Nineveh depends on only two parameters: the cycle number c of the year in question within the 19-year cycle and the weekday t of its first day. Instead of c one can, of course, also use the equivalent parameter e, the "epact" of the year.

The simplest way of obtaining, from a given pair of parameters c (or e) and t, the value of n consists in computing the date f of Easter and going back 69 days to n on the basis of (2).

The following Table XII illustrates this procedure by systematically following the method explained above, p. 45, for finding f and from it n. It would have been equally possible to go directly from pw to n but the explicit use of f makes the causal connections clearer.

Table XII

c	t	e	m = 30−e	p = m+190	mw	pw = mw+1	f	n = f−69
1	1	0	I 30	VIII 10	2	3	VIII 12	VI 3
	2				3	4	11	2
	3				4	5	17	8
	4				5	6	16	7
	5				6	7	15	6
	6				7	1	14	5
	7				1	2	13	4
2	1	11	I 19	VII 29	5	6	VIII 5	V 26
	2				6	7	4	25
	3				7	1	3	24
	4				1	2	2	23

etc.

The final result for all $19 \cdot 7 = 133$ combinations of c and t is shown in Table XIII. Since $532 = 4 \cdot 133$ these values will be repeated four times in each 532-year table.

Table XIII

c	t							e
	1	2	3	4	5	6	7	
1	3	2	8	7	6	5	4	0
2	26	25	24	23	22	21	27	11
3	10	16	15	14	13	12	11	22
4	3	2	1	30	29	5	4	3
5	19	18	24	23	22	21	20	14
6	10	9	8	7	13	12	11	25
7	26	2	1	30	29	28	27	6
8	17	16	15	21	20	19	18	17
9	10	9	8	7	6	5	4	28
10	26	25	24	23	29	28	27	9
11	17	16	15	14	13	12	18	20
12	3	2	1	7	6	5	4	1
13	26	25	24	23	22	21	20	12
14	10	9	15	14	13	12	11	23
15	3	2	1	30	29	28	4	4
16	19	18	17	23	22	21	20	15
17	10	9	8	7	6	12	11	26
18	26	25	1	30	29	28	27	7
19	17	16	15	14	20	19	18	18

A table exactly the type of our Table XIII is preserved in the zīj-i mufrad by Ṭabarī (around 1230, in Persian)[68]. Its purpose is to find for

[68] Cf. Saliba [1970] p. 208, Fig. 7 and p. 180 (VIII), p. 194 (K).

given c and t the date of Lent (bf') which has a given distance from n:

$$bf' = n + 21 \equiv n - 9 \bmod 30$$

as compared with our $bf = n + 14$[69]. Furthermore, the epoch for the 19-year cycles has been changed so that $c = 1$ in our table corresponds to $c = 6$ in the Persian table. Adding 9 to the numbers of n shown in Table XIII gives the numbers found in Ṭabarī's table (excepting many scribal errors).

ν. *Lunar Date of I 1*

Cf. s. v. Maṭqeʿe and Epact, p. 190.

p. *Date of Passover*

The lunar date $\bar{p}$ of Passover is by definition 14. In contrast to fāsikā for Easter Sunday, Passover is called feśḥ, though both terms are eventually derived from the Hebrew פסח.

The date p is related to the maṭqeʿe m, the Jewish New Years Day, by

(1) $$p = m + 190 \equiv m + 10 \bmod 30$$

and therefore (since $e + m = 30$) to the epact by

(2) $$e + p \equiv 10 \bmod 30.$$

Since e repeats its values after 19 years, the same also holds for p, because of (2).

We have special tables[70] that give the dates of Passover for 28 19-year cycles as function of the weekday (j) of the civil New Years Day, Maskaram 1. For example,

6	7	1	3	4		6	7
10	29	18	7	26		3	22
					etc.		
M.	Mr.	L.	J.	M.		Mr.	L.
39							

The first line gives the weekdays j for the cycle years from $k = 39$ (cf. the 4th line) to 57[71]. The second line repeats in each of the 28 tables the dates p from $c = 1$ to 19, while the third line corresponds to the sequence

[69] Cf. above p. 49 (3).
[70] EMML 2077 156ᵃ—157ᵇ but copied in incorrect order.
[71] Actually for $k = 457$ to 475 (table 25); cf. the preceding note.

of the weekdays in their dependence on the Alexandrian intercalation pattern. Such a table is nothing but an excerpt of 3 columns from the 532-year tables.

pw. Weekday of Passover

Some of our tables contain a special column for the weekday *pw* of Passover (in the norm Sunday = 1), others combine in the same column *p* and *pw*. Since $p = m + 190$, it is easy to find *pw* from *mw*:

(1) $$pw \equiv mw + 1 \mod 7$$

and therefore with (2) from p. 48

(2) $$pw \equiv \begin{cases} j+m \\ j+m+2 \end{cases} \mod 7 \begin{cases} \text{if } 15 \leq m \leq 30 \\ \text{if } \ 1 \leq m \leq 13. \end{cases}$$

A simple method of checking can make use of the fact that the feria of *f*, Easter Sunday, is 8. Thus $pw = 8 - (f - p)$; but $f - p = \bar{f} - \bar{p}$ and $\bar{p} = 14$, therefore

(3) $$pw + \bar{f} = 22.$$

This relationship is useful for tables which give the "luna" $\bar{f}$ of Easter Sunday.

t. Ṭentyon

Weekday of Maskaram 1, expressed numerically with the norm 1 = Wednesday. Changing to the norm Sunday = 1 means replacing *t* by the "Day of John" (*j*) so that

(1) $$t \equiv j - 3 \mod 7.$$

For a year given in the era W or G the ṭentyon can be found from

$$t \equiv W + [W/4] - 1 \equiv G + [G/4] - 1 \mod 7.$$

Cf. s. v. Ṭentyon and Day of John, p. 221.

tb. Tabernacle

Jewish feast, called maṣalat, "Tabernacle". In the schematic lunar calendar, referred to by "evening epoch", its date is 15[72]. Therefore the column *tb* in the 532-year tables is followed by a column $\overline{tb} = 15$.

In relation to the maṭqeʿe *tb* is given by

(1) $$tb \equiv m + 14 \mod 30.$$

tw. Weekday of t

[72] Vat 119 86ᵇ II, 15—17.

The 532-year tables give also in words the weekday which is represented by t. This information is always combined in the same column with the name of the Evangelist of the year (ε).

Some chronological lists use more explicit formulations[73], e. g., "the beginning of Maskaram (fell) on the day . . ."

yk. Yom Kippur

The Jewish feast of "Yom Kippur" or "Atonement" is called ṣoma 'ayehud, "Jewish Fast" or badr. Its civil date is obtainable from

$$yk = m + 9. \quad (1)$$

The lunar date $\overline{yk}$ of Yom Kippur is 10.

4. Concordances with other Calendars

In many of our manuscripts we find lists which equate the months of the Ethiopic calendar with the months in some foreign system[74]. These concordances are usually quite schematic, ignoring, e. g., differences in the norm for the lengths of individual months. Also the origin of foreign calendars is usually incorrectly described, e. g. calling "Hebrew" or "Jewish" what is actually the Syriac calendar[75], or "Arabic" (as distinct from "Coptic") what is in fact only the Arabic version of the Coptic month-names[76]. The "Roman" calendar may be the Latin julian calendar or again the Syriac calendar.

It is clear that the Ethiopic material cannot compete in any way with Islamic chronological works, e. g., Birūnī's "Chronology".

Arabic Calendar

Abū Šākir's work contains also tables[77] that allow one to find for 12 consecutive 19-year cycles the month of the Arabic lunar calendar that corresponds (in its major part) to the Ethiopic month Maskaram. The number $12 \cdot 19 = 228$ civil years is chosen because it represents the period for the numbering of the Arabic months.

The principle upon which this concordance is erected is shown in the following table which concerns the first two 19-year cycles.

---

73 CSCO 61/66, pp. 26—87/27—94 passim.

74 E. g., comparison with "Roman" and "Coptic" months in Codices Orient. Bibl. reg. Hafniensis, pars 3, p. 79, cod. II, 3 fol. 86^a (Dillmann).

75 E. g., Vindob 6 1^a, 1 to 2^a, 21; 3^a, 10 to 15.

76 E. g., Princ 5884 12^a II, 12 to 12^b II, 3.

77 BM Add 16252 161^a—163^b (very disorderly written).

civil year:	1	2	3	4	5	6	7	8	9	10	11	12	13	14	15	16	17	18	19	
lunar year:	12	12	13	12	12	13	12	12	13	12	13	12	12	13	12	12	13	12	13	
from 1	I	I	I	II	II	II	III	III	III	IV	IV	V	V	V	VI	VI	VI	VII	VII	to 19
20	VIII	VIII	VIII	IX	IX	IX	X	X	X	XI	XI	XII	XII	XII	I	I	I	II	II	38
39	III	III	III	IV		etc.														

The horizontal entries are (a) civil years and (b) lunar years where "13" indicates an intercalary year with 13 lunar months (thus 7 times in 19 years). The Roman numerals represent the Arabic months (I $\approx$ Muḥarram). This pattern repeats itself after the twelfth cycle (from 210 to 228).

Dates according to the "Hebrews"

The "Didascalia" contains some alternative dates "according to the Hebrews":

Harden, Didasc. p. 122:	Ethiopic	IV 29	Hebrews	IX 25
		V 11		X 6
p. 131:		VII 25		VII 22

The month names in the first two dates could refer to a year beginning with Nisan. The differences +4, +5, +3 between the day numbers are perhaps caused by the differences between lunar dates (Hebrews) and Alexandrian civil dates.

5. *Examples*

1. "W = 7178, year of Mark, solar epact 4, lunar epact 26"[78]. In checking these data one finds

$$7178 \equiv 2 \quad \text{mod } 4 \qquad \text{thus } \varepsilon = 2 \text{ i. e. Mark}$$

$$W - 1 = 7177$$

$$[W/4] = 1794$$

$$t = 8971 \equiv 4 \text{ mod } 7 \qquad \text{"solar epact"} = \text{ṭentyon.}$$

This term apparently suggested calling the next parameter "lunar epact" instead of maṭqeʿe. In fact

$$7178 \equiv 15 \text{ mod } 19 \qquad \text{hence } c = 15$$

$$m = (c - 1) \cdot 19 = 266 \equiv 26 \text{ mod } 30;$$

for the lunar epact one has $e = 30 - m = 4$.

2. Strelcyn, Cat. John Rylands, in the colophon to No. 1, written in the year W 7147 (= J 1647 = G 1295), adds in line 14 a "sic" to

[78] Dillmann, Bodl. Catal. No. 16, p. 29 (for the preceding text, cf. s. v. Eras, p. 120).

"lunar epact 22" although 22 is the correct number for e, just as the preceding "solar epact 7" is for t. Indeed, G 1295 $\equiv$ 231 mod 532 and to $k = 231$ belongs in the 532-year tables $e = 22$, $t = 7$. This allows us to restore in line 11 for the cycle number [2] 31. The weekday as defined by $t = 7$ is Tuesday, not Wednesday as stated on p. 1.

3. In a discussion about the year 7000 we find the following statement[79]: "For this year, which has an epagomenal day 6, the indictio is 11, the ṭentyon 2, the epact 17, the solar epact 2".
Indeed:

W 7000 — 6916 = 84 (in the cycle G_3) and $k = 84 \equiv 0 = \varepsilon$ mod 4. Hence W 7000 is an intercalary year[80].

For the indictio we have $i = 0$ for the year D 13[81]. Thus $i = 11$ is correct for D 39 = W 5815 $\equiv$ W 7000 mod 15.

For $k = 84$ the 532-year tables give $j = 2$ (not t) and, correctly, $e = 17$.

Finally, the text describes the 28-year solar cycle and correctly states that 7000 is divisible by 28. The scribe, however, confused solar cycle with solar epact, which is another term for ṭentyon[82], and thus repeated the number $j = 2$ (for t).

6. *Bibliographical Notes.*

The first discussion of the technical aspects of the Ethiopic calendar we owe to Scaliger in his famous work *De emendatione temporum* (editions 1583, 1598, 1629). The Ethiopic sections seem to be based on information obtained from a native, living in Rome, who had only a very limited amount of relevant texts at his disposal.

Modern chronological literature has contributed much toward obscuring the simplicity and methodological elegance of the Ethiopic(-Alexandrian)-Easter computus by contaminating it with the Roman-julian (or even Gregorian) calendar. A result of this scholarly provincialism is such clumsy tables as those given in the works of Chaine (1925), Grumel (1960), and others which, by their sheer massiveness, obstruct access to the simple ideas in the primary sources[83]. Ginzel (Hdb. III, 1914,

[79] Berol 84 21ª II, 1 to 21ᵇ I, 5.

[80] Cf. s. v. Eras p. 117 Table I, and above p. 30.

[81] Cf. s. v. Eras p. 120.

[82] Cf. s. v. Ṭentyon p. 219.

[83] Also, the mathematical level of these works is more reminiscent of Beda Venerabilis than of Scaliger. Grumel, e. g., is of the opinion that the years +1 and −1 are contiguous (cf. Chron. Byz., table p. 30 et passim). The lack of indices makes Grumel's handbook even more useless.

pp. 324—326) depended on secondary sources and transmitted, e. g., incorrect interpretations on *m* and *t*. The only competent edition of a chronological treatise I know of is by Mauro da Leonessa (1943)[84]. Weld Blundell, Roy. Chron. (1922) also gave a reasonable summary of chronological matters[85]. Grébaut in his innumerable notes dealt only with trivial sections of calendaric treatises.

For the history of the Alexandrian Easter computus and its connections with Western medieval procedures, cf. the fundamental studies by Ed. Schwartz, Ostertafeln (1905) and Osterbetrachtungen (1906). Others are Krusch, Chron. (1880) and Rühl, Chron. (1897). The Dissertazioni Cronologici by Mauro da Leonessa (1934) are based on a practical knowledge of the contemporary usage and tradition of the calendar. Since many of the theoretical statements are incorrect or are lacking precision this work should only be used with caution.

Appendix I. The 532-year Table

Incredible as it may seem, the 532-year table, which lies at the center of all Ethiopic calendaric procedures, has never before been published. Its existence was known to European scholarship for well over a hundred years[86], but its numerical contents were ignored and chronological experts like Wüstenfeld and Mahler, Chaine, and Grumel did not make any attempt to grasp its structure. In consequence of this neglect chronological data in colophons and incipits of texts reproduced in the catalogues of the European collections abound with errors which could easily have been avoided by a simple comparison with the entries in the 532-year tables. Nor did scholars concerned with the Easter controversies between the Alexandrian and the "Latin" Computus realize that the Alexandrian version is fully available in Ethiopic tables, carefully stored away in our museums and libraries.

The complete tables, for 28 19-year cycles, are published below, pp. 59—63. To avoid unnecessary duplication, I have collected all columns that have exactly a 19-year period into one initial table headed

---

[84] From Vat 61.

[85] Roy. Chron., pp. 493—505, though with some errors.

[86] The headings of its 20-odd columns are listed in the catalogues of Wright, B. M. Cat. p. 227 (1877); Rhodokanakis (1906) p. 82; Grébaut-Tisserant, Cod. Vat. p. 484 (1935); Hammerschmidt, Ṭānā I, p. 157 f. (1973). A short excerpt of the tables themselves (for $k = 1$ to 9 and 19 to 21) from BM 815, was published by Weld Blundell, Roy. Chron., p. 502 f. (with scribal errors in the dates for *p* and *f*, at $c = 9$).

"XIX". The cycle number c suffices to associate each line of table "XIX" with the corresponding line in any one of the remaining 28 tables. Entirely omitted are four columns which display the same number in all tables, representing the lunar dates for the Jewish New Years Day ($\overline{m} = 1$), for Yom Kippur ($\overline{yk} = 10$), Tabernacle ($\overline{tb} = 15$), and Passover ($\overline{p} = 14$). Not all 532-year tables contain the weekdays for maṭqe'e (mw) and Passover (pw). Since $mw \equiv pw - 1 \pmod 7$ I have tabulated only pw. Also omitted are the words for the weekdays of Maskaram 1 because exactly the same information is available in numerical form in the columns j (Sunday = 1) and t (Wednesday = 1). The actual arrangement in the texts can be seen in our Table III on p. 31.

The numbers in all tables can easily be checked. In Table "XIX" all data depend on the epact e which increases, modulo 30, by 11 from year to year, while le, m, yk, tb, and p decrease by the same amount. Furthermore, one has the following relations:

$$e + m \equiv 30 \quad le \equiv m + 6 \quad yk \equiv m + 9 \quad tb \equiv m + 14 \quad p \equiv m + 10 \mod 30.$$

The columns ep, ε, j and t reflect the 4-year Alexandrian intercalation cycle and

$$t \equiv j + 4 \qquad \text{or} \qquad j \equiv t + 3 \mod 7.$$

For the determination of n from t and c, cf. p. 50 Table XIII. From n one obtains

$$bf \equiv n + 14 \qquad f \equiv bf - 5 \mod 30.$$

Finally,

$$pw \equiv \begin{cases} j + m \\ j + m + 2 \end{cases} \mod 7 \quad \begin{cases} \text{if } 15 \leqq m \leqq 30 \\ \phantom{\text{if }} 1 \leqq m \leqq 13. \end{cases}$$

All manuscripts that display these tables contain innumerable scribal errors.

The numbers in the columns m, yk, tb, n, bf, p, and f are dates which belong to specific months in the civil calendar according to the following rules[87]:

$$\text{I } 15 \leqq m \leqq \text{II } 13$$
$$\text{I } 24 \leqq yk \leqq \text{II } 22$$
$$\text{II } 29 \leqq tb \leqq \text{III } 27$$
$$\text{VII } 25 \leqq p \leqq \text{VIII } 23$$

$$\text{V } 17 \leqq n \leqq \text{VI } 21$$
$$\text{VI } 1 \leqq bf \leqq \text{VII } 5$$
$$\text{VII } 26 \leqq f \leqq \text{VIII } 30$$

[87] Cf. s. v. Feasts p. 133.

For the Jewish festivals these boundaries uniquely determine the month to which a given day number belongs. For Christian feasts there remains an ambiguity for four dates which can belong in principle to either one of two consecutive months. The decision depends on the ṭentyon, the weekday of Maskaram 1. It suffices to give the rule for Easter[88] which normally falls in month VIII, with the following exceptions:

f falls in VII	30 only when $t =$	6
	29	7
	28	1
	27	2
	26	3.

It is easy to check these rules in the tables. Dates in column f which belong to month VII (Magābit) are marked in the texts by the letter gā. In the tables here I have underlined these numbers. In the case of Passover, no ambiguity exists; the majority of dates belongs to month VIII (Miyāzyā) but four dates (between 25 and 30) belong to VII. For the sake of convenience I have also underlined these dates in Table "XIX".

The cycle numbers k count the years from 1 to 532 in each of the 532-year cycles of the eras W and G[89]. If one is dealing with a text based on the era D one has to add 76 to the numbers k found in our tables. The cycle number c remains valid for all three eras. For the era J, however, not only has k to be increased by 180 (mod 532) but also c is replaced by $c - 10$ (mod 19) in order to use tables based on the era G.

Ordinarily the 532-year tables begin with $k = 1$ of the era G, but we also have arrangements (of tables for 6 to 8 parameters only) for which $k = 153$ is the first year in the first 19-year cycle tabulated[90]. This shift of the starting point is expressly noted in Berol 994 where fol. 69^a to 73^b cover the years from $k = 153$ to 532 and fol. 74^a to 75^b the years from $k = 1$ (ṭenta qamar) to $k = 152$. One text[91] covers only the 7 cycles from $k = 153$ to 285. The year $k = 153$ is the only year in a 532-year cycle which is not only the first year of a 19-year cycle but also of a 28-year solar cycle[92].

---

[88] Cf. s. v. Feasts p. 147 (Easter).

[89] Cf. s. v. Eras p. 117.

[90] BM Add 16220 1^a—5^b I; BM Add 19658 51^a—57^b; Princ 5884 1^a—5^b I.

[91] BN Abb 223 1^a—6^a.

[92] Cf. below s. v. Cycles p. 88.

XIX

c	le	e	m	yk	tb	p
1	0	0	30	9	14	10
2	25	11	19	28	3	29
3	14	22	8	17	22	18
4	3	3	27	6	11	7
5	22	14	16	25	30	26
6	11	25	5	14	19	15
7	30	6	24	3	8	4
8	19	17	13	22	27	23
9	8	28	2	11	16	12
10	27	9	21	30	5	1
11	16	20	10	19	24	20
12	5	1	29	8	13	9
13	24	12	18	27	2	28
14	13	23	7	16	21	17
15	2	4	26	5	10	6
16	21	15	15	24	29	25
17	10	26	4	13	18	14
18	29	7	23	2	7	3
19	18	18	12	21	26	22

1

k	c	ep	ϵ	j	t	n	bf	pw	f
1	1	5	M.	3	7	4	18	5	13
2	2	5	Mr.	4	1	26	10	2	5
3	3	5	L.	5	2	16	30	1	25
4	4	6	J.	7	4	30	14	6	9
5	5	5	M.	1	5	22	6	3	1
6	6	5	Mr.	2	6	12	26	2	21
7	7	5	L.	3	7	27	11	6	6
8	8	6	J.	5	2	16	30	6	25
9	9	5	M.	6	3	8	22	3	17
10	10	5	Mr.	7	4	23	7	7	2
11	11	5	L.	1	5	13	27	6	22
12	12	6	J.	3	7	4	18	4	13
13	13	5	M.	4	1	26	10	1	5
14	14	5	Mr.	5	2	9	23	7	18
15	15	5	L.	6	3	1	15	4	10
16	16	6	J.	1	5	22	6	2	1
17	17	5	M.	2	6	12	26	1	21
18	18	5	Mr.	3	7	27	11	5	6
19	19	5	L.	4	1	17	1	4	26

2

k	c	ep	ϵ	j	t	n	bf	pw	f
20	1	6	J.	6	3	8	22	1	7
21	2	5	M.	7	4	23	7	5	2
22	3	5	Mr.	1	5	13	27	4	22
23	4	5	L.	2	6	5	19	1	14
24	5	6	J.	4	1	19	3	6	23
25	6	5	M.	5	2	9	23	5	13
26	7	5	Mr.	6	3	1	15	2	10
27	8	5	L.	7	4	21	5	1	30
28	9	6	J.	2	6	5	19	6	14
29	10	5	M.	3	7	27	11	3	6
30	11	5	Mr.	4	1	17	1	2	26
31	12	5	L.	5	2	2	16	6	11
32	13	6	J.	7	4	23	7	4	2
33	14	5	M.	1	5	13	27	3	22
34	15	5	Mr.	2	6	28	12	7	7
35	16	5	L.	3	7	20	4	4	29
36	17	6	J.	5	2	9	23	4	18
37	18	5	M.	6	3	1	15	1	10
38	19	5	Mr.	7	4	14	28	7	23

3

k	c	ep	ϵ	j	t	n	bf	pw	f
39	1	5	L.	1	5	6	20	3	15
40	2	6	J.	3	7	27	11	1	6
41	3	5	M.	4	1	10	24	7	19
42	4	5	Mr.	5	2	2	16	4	11
43	5	5	L.	6	3	24	8	1	3
44	6	6	J.	1	5	13	27	1	22
45	7	5	M.	2	6	28	12	5	7
46	8	5	Mr.	3	7	18	2	4	27
47	9	5	L.	4	1	10	24	1	19
48	10	6	J.	6	3	24	8	6	3
49	11	5	M.	7	4	14	28	5	23
50	12	5	Mr.	1	5	6	20	2	15
51	13	5	L.	2	6	21	5	6	30
52	14	6	J.	4	1	10	24	6	19
53	15	5	M.	5	2	2	16	3	11
54	16	5	Mr.	6	3	17	1	7	26
55	17	5	L.	7	4	7	21	6	16
56	18	6	J.	2	6	28	12	4	7
57	19	5	M.	3	7	18	2	3	27

4

k	c	ep	ϵ	j	t	n	bf	pw	f
58	1	5	Mr.	4	1	3	17	6	12
59	2	5	L.	5	2	25	9	3	4
60	3	6	J.	7	4	14	28	3	23
61	4	5	M.	1	5	29	13	7	8
62	5	5	Mr.	2	6	21	5	4	30
63	6	5	L.	3	7	11	25	3	20
64	7	6	J.	5	2	2	16	1	11
65	8	5	M.	6	3	15	29	7	24
66	9	5	Mr.	7	4	7	21	4	16
67	10	5	L.	1	5	29	13	1	8
68	11	6	J.	3	7	18	2	1	27
69	12	5	M.	4	1	3	17	5	12
70	13	5	Mr.	5	2	25	9	2	4
71	14	5	L.	6	3	15	29	1	24
72	15	6	J.	1	5	29	13	6	8
73	16	5	M.	2	6	21	5	3	30
74	17	5	Mr.	3	7	11	25	3	20
75	18	5	L.	4	1	26	10	6	5
76	19	6	J.	6	3	15	29	6	24

5

k	c	ep	ϵ	j	t	n	bf	pw	f
77	1	5	M.	7	4	7	21	2	16
78	2	5	Mr.	1	5	22	6	6	1
79	3	5	L.	2	6	12	26	5	21
80	4	6	J.	4	1	3	17	3	12
81	5	5	M.	5	2	18	2	7	27
82	6	5	Mr.	6	3	8	22	6	17
83	7	5	L.	7	4	30	14	3	9
84	8	6	J.	2	6	19	3	3	28
85	9	5	M.	3	7	4	18	7	13
86	10	5	Mr.	4	1	26	10	4	5
87	11	5	L.	5	2	16	30	3	25
88	12	6	J.	7	4	7	21	1	16
89	13	5	M.	1	5	22	6	5	1
90	14	5	Mr.	2	6	12	26	4	21
91	15	5	L.	3	7	4	18	1	13
92	16	6	J.	5	2	18	2	6	27
93	17	5	M.	6	3	8	22	5	17
94	18	5	Mr.	7	4	30	14	2	9
95	19	5	L.	1	5	20	4	1	29

6

k	c	ep	∈	j	t	n	bf	pw	f
96	1	6	J.	3	7	4	18	5	13
97	2	5	M.	4	1	26	10	2	5
98	3	5	Mr.	5	2	16	30	1	25
99	4	5	L.	6	3	1	15	5	10
100	5	6	J.	1	5	22	6	3	1
101	6	5	M.	2	6	12	26	2	21
102	7	5	Mr.	3	7	27	11	6	6
103	8	5	L.	4	1	17	1	5	26
104	9	6	J.	6	3	8	22	3	17
105	10	5	M.	7	4	23	7	7	2
106	11	5	Mr.	1	5	13	27	6	22
107	12	5	L.	2	6	5	19	3	14
108	13	6	J.	4	1	26	10	1	5
109	14	5	M.	5	2	9	23	7	18
110	15	5	Mr.	6	3	1	15	4	10
111	16	5	L.	7	4	23	7	1	2
112	17	6	J.	2	6	12	26	1	21
113	18	5	M.	3	7	27	11	5	6
114	19	5	Mr.	4	1	17	1	4	26

7

k	c	ep	∈	j	t	n	bf	pw	f
115	1	5	L.	5	2	2	16	7	11
116	2	6	J.	7	4	23	7	5	2
117	3	5	M.	1	5	13	27	4	22
118	4	5	Mr.	2	6	5	19	1	14
119	5	5	L.	3	7	20	4	5	29
120	6	6	J.	5	2	9	23	5	18
121	7	5	M.	6	3	1	15	2	10
122	8	5	Mr.	7	4	21	5	1	30
123	9	5	L.	1	5	6	20	5	15
124	10	6	J.	3	7	27	11	3	6
125	11	5	M.	4	1	17	1	2	26
126	12	5	Mr.	5	2	2	16	6	11
127	13	5	L.	6	3	24	8	3	3
128	14	6	J.	1	5	13	27	3	22
129	15	5	M.	2	6	28	12	7	7
130	16	5	Mr.	3	7	20	4	4	29
131	17	5	L.	4	1	10	24	3	19
132	18	6	J.	6	3	1	15	1	10
133	19	5	M.	7	4	14	28	7	23

8

k	c	ep	∈	j	t	n	bf	pw	f
134	1	5	Mr.	1	5	6	20	3	15
135	2	5	L.	2	6	21	5	7	30
136	3	6	J.	4	1	10	24	7	19
137	4	5	M.	5	2	2	16	4	11
138	5	5	Mr.	6	3	24	8	1	3
139	6	5	L.	7	4	7	21	7	16
140	7	6	J.	2	6	28	12	5	7
141	8	5	M.	3	7	18	2	4	27
142	9	5	Mr.	4	1	10	24	1	19
143	10	5	L.	5	2	25	9	5	4
144	11	6	J.	7	4	14	28	5	23
145	12	5	M.	1	5	6	20	2	15
146	13	5	Mr.	2	6	21	5	6	30
147	14	5	L.	3	7	11	25	5	20
148	15	6	J.	5	2	2	16	3	11
149	16	5	M.	6	3	17	1	7	26
150	17	5	Mr.	7	4	7	21	6	16
151	18	5	L.	1	5	29	13	3	8
152	19	6	J.	3	7	18	2	3	27

9

k	c	ep	∈	j	t	n	bf	pw	f
153	1	5	M	4	1	3	17	6	12
154	2	5	Mr.	5	2	25	9	3	4
155	3	5	L.	6	3	15	29	2	24
156	4	6	J	1	5	29	13	7	8
157	5	5	M.	2	6	21	5	4	30
158	6	5	Mr.	3	7	11	25	3	20
159	7	5	L.	4	1	26	10	7	5
160	8	6	J.	6	3	15	29	7	24
161	9	5	M.	7	4	7	21	4	16
162	10	5	Mr.	1	5	29	13	1	8
163	11	5	L.	2	6	12	26	7	21
164	12	6	J.	4	1	3	17	5	12
165	13	5	M.	5	2	25	9	2	4
166	14	5	Mr.	6	3	15	29	1	24
167	15	5	L.	7	4	30	14	5	9
168	16	6	J.	2	6	21	5	3	30
169	17	5	M.	3	7	11	25	2	20
170	18	5	Mr.	4	1	26	10	6	5
171	19	5	L.	5	2	16	30	5	25

10

k	c	ep	∈	j	t	n	bf	pw	f
172	1	6	J.	7	4	7	21	2	16
173	2	5	M.	1	5	22	6	6	1
174	3	5	Mr.	2	6	12	26	5	21
175	4	5	L.	3	7	4	18	2	13
176	5	6	J.	5	2	18	2	7	27
177	6	5	M.	6	3	8	22	6	17
178	7	5	Mr.	7	4	30	14	3	9
179	8	5	L.	1	5	20	4	2	29
180	9	6	J.	3	7	4	18	7	13
181	10	5	M.	4	1	26	10	4	5
182	11	5	Mr.	5	2	16	30	3	25
183	12	5	L.	6	3	1	15	7	10
184	13	6	J.	1	5	22	6	5	1
185	14	5	M.	2	6	12	26	4	21
186	15	5	Mr.	3	7	4	18	1	13
187	16	5	L.	4	1	19	3	5	28
188	17	6	J.	6	3	8	22	5	17
189	18	5	M.	7	4	30	14	2	9
190	19	5	Mr.	1	5	20	4	1	29

11

k	c	ep	∈	j	t	n	bf	pw	f
191	1	5	L.	2	6	5	19	4	14
192	2	6	J.	4	1	26	10	2	5
193	3	5	M.	5	2	16	30	1	25
194	4	5	Mr.	6	3	1	15	5	10
195	5	5	L.	7	4	23	7	2	2
196	6	6	J.	2	6	12	26	2	21
197	7	5	M.	3	7	27	11	6	6
198	8	5	Mr.	4	1	17	1	5	26
199	9	5	L.	5	2	9	23	2	18
200	10	6	J.	7	4	23	7	7	2
201	11	5	M.	1	5	13	27	6	22
202	12	5	Mr.	2	6	5	19	3	14
203	13	5	L.	3	7	20	4	7	29
204	14	6	J.	5	2	9	23	7	18
205	15	5	M.	6	3	1	15	4	10
206	16	5	Mr.	7	4	23	7	1	2
207	17	5	L.	1	5	6	20	7	15
208	18	6	J.	3	7	27	11	5	6
209	19	5	M.	4	1	17	1	4	26

12

k	c	ep	ϵ	j	t	n	bf	pw	f
210	1	5	Mr.	5	2	2	16	7	11
211	2	5	L.	6	3	24	8	4	3
212	3	6	J.	1	5	13	27	4	22
213	4	5	M.	2	6	5	19	1	14
214	5	5	Mr	3	7	20	4	5	29
215	6	5	L.	4	1	10	24	4	19
216	7	6	J.	6	3	1	15	2	10
217	8	5	M.	7	4	21	5	1	30
218	9	5	Mr.	1	5	6	20	5	15
219	10	5	L.	2	6	28	12	2	7
220	11	6	J.	4	1	17	1	2	26
221	12.	5	M.	5	2	2	16	6	11
222	13	5	Mr.	6	3	24	8	3	3
223	14	5	L.	7	4	14	28	2	23
224	15	6	J.	2	6	28	12	7	7
225	16	5	M.	3	7	20	4	4	29
226	17	5	Mr.	4	1	10	24	3	19
227	18	5	L.	5	2	25	9	7	4
228	19	6	J	7	4	14	28	7	23

13

k	c	ep	ϵ	j	t	n	bf	pw	f
229	1	5	M.	1	5	6	20	3	15
230	2	5	Mr.	2	6	21	5	7	30
231	3	5	L.	3	7	11	25	6	20
232	4	6	J.	5	2	2	16	4	11
233	5	5	M.	6	3	24	8	1	3
234	6	5	Mr.	7	4	7	21	7	16
235	7	5	L.	1	5	29	13	4	8
236	8	6	J.	3	7	18	2	4	27
237	9	5	M.	4	1	10	24	1	19
238	10	5	Mr.	5	2	25	9	5	4
239	11	5	L.	6	3	15	29	4	24
240	12	6	J.	1	5	6	20	2	15
241	13	5	M.	2	6	21	5	6	30
242	14	5	Mr.	3	7	11	25	5	20
243	15	5	L.	4	1	3	17	2	12
244	16	6	J.	6	3	17	1	7	26
245	17	5	M.	7	4	7	21	6	16
246	18	5	Mr.	1	5	29	13	3	8
247	19	5	L	2	6	19	3	2	28

14

k	c	ep	ϵ	j	t	n	bf	pw	f
248	1	6	J.	4	1	3	17	6	12
249	2	5	M.	5	2	25	9	3	4
250	3	5	Mr.	6	3	15	29	2	24
251	4	5	L.	7	4	30	14	6	9
252	5	6	J.	2	6	21	5	4	30
253	6	5	M.	3	7	11	25	3	20
254	7	5	Mr.	4	1	26	10	7	5
255	8	5	L.	5	2	16	30	6	25
256	9	6	J.	7	4	7	21	4	16
257	10	5	M.	1	5	29	13	1	8
258	11	5	Mr.	2	6	12	26	7	21
259	12	5	L.	3	7	4	18	4	13
260	13	6	J.	5	2	25	9	2	4
261	14	5	M.	6	3	15	29	1	24
262	15	5	Mr.	7	4	30	14	5	9
263	16	5	L.	1	5	22	6	2	1
264	17	6	J.	3	7	11	25	2	20
265	18	5	M.	4	1	26	10	6	5
266	19	5	Mr.	5	2	16	30	5	25

15

k	c	ep	ϵ	j	t	n	bf	pw	f
267	1	5	L.	6	3	8	22	1	17
268	2	6	J.	1	5	22	6	6	1
269	3	5	M.	2	6	12	26	5	21
270	4	5	Mr.	3	7	4	18	2	13
271	5	5	L.	4	1	19	3	6	28
272	6	6	J.	6	3	8	22	6	17
273	7	5	M.	7	4	30	14	3	9
274	8	5	Mr.	1	5	20	4	2	29
275	9	5	L.	2	6	5	19	6	14
276	10	6	J.	4	1	26	10	4	5
277	11	5	M.	5	2	16	30	3	25
278	12	5	Mr.	6	3	1	15	7	10
279	13	5	L.	7	4	23	7	4	2
280	14	6	J.	2	6	12	26	4	21
281	15	5	M.	3	7	4	18	1	13
282	16	5	Mr.	4	1	19	3	5	28
283	17	5	L.	5	2	9	23	4	18
284	18	6	J.	7	4	30	14	2	9
285	19	5	M.	1	5	20	4	1	29

16

k	c	ep	e	j	t	n	bf	pw	f
286	1	5	Mr.	2	6	5	19	4	14
287	2	5	L.	3	7	27	11	1	6
288	3	6	J.	5	2	16	30	1	25
289	4	5	M.	6	3	1	15	5	10
290	5	5	Mr.	7	4	23	7	2	2
291	6	5	L.	1	5	13	27	1	22
292	7	6	J.	3	7	27	11	6	6
293	8	5	M.	4	1	17	1	5	26
294	9	5	Mr.	5	2	9	23	2	18
295	10	5	L.	6	3	24	8	6	3
296	11	6	J.	1	5	13	27	6	22
297	12	5	M.	2	6	5	19	3	14
298	13	5	Mr.	3	7	20	4	7	29
299	14	5	L.	4	1	10	24	6	19
300	15	6	J.	6	3	1	15	4	10
301	16	5	M.	7	4	23	7	1	2
302	17	5	Mr.	1	5	6	20	7	15
303	18	5	L	2	6	28	12	4	7
304	19	6	J	4	1	17	1	4	26

17

k	c	ep	ϵ	j	t	n	bf	pw	f
305	1	5	M.	5	2	2	16	7	11
306	2	5	Mr.	6	3	24	8	4	3
307	3	5	L.	7	4	14	28	3	23
308	4	6	J.	2	6	5	19	1	14
309	5	5	M.	3	7	20	4	5	29
310	6	5	Mr.	4	1	10	24	4	19
311	7	5	L.	5	2	2	16	1	11
312	8	6	J.	7	4	21	5	1	30
313	9	5	M.	1	5	6	20	5	15
314	10	5	Mr.	2	6	28	12	2	7
315	11	5	L.	3	7	18	2	1	27
316	12	6	J.	5	2	2	16	6	11
317	13	5	M.	6	3	24	8	3	3
318	14	5	Mr.	7	4	14	28	2	23
319	15	5	L.	1	5	29	13	6	8
320	16	6	J.	3	7	20	4	4	29
321	17	5	M.	4	1	10	24	3	19
322	18	5	Mr.	5	2	25	9	7	4
323	19	5	L.	6	3	15	29	6	24

18

k	c	ep	ε	j	t	n	bf	pw	f
324	1	6	J.	1	5	6	20	3	15
325	2	5	M.	2	6	21	5	7	30
326	3	5	Mr.	3	7	11	25	6	20
327	4	5	L.	4	1	3	17	3	12
328	5	6	J.	6	3	24	8	1	3
329	6	5	M.	7	4	7	21	7	16
330	7	5	Mr.	1	5	29	13	4	8
331	8	5	L.	2	6	19	3	3	28
332	9	6	J.	4	1	10	24	1	19
333	10	5	M.	5	2	25	9	5	4
334	11	5	Mr.	6	3	15	29	4	24
335	12	5	L.	7	4	7	21	1	16
336	13	6	J.	2	6	21	5	6	30
337	14	5	M.	3	7	11	25	5	20
338	15	5	Mr.	4	1	3	17	2	12
339	16	5	L.	5	2	18	2	6	27
340	17	6	J.	7	4	7	21	6	16
341	18	5	M.	1	5	29	13	3	8
342	19	5	Mr.	2	6	19	3	2	28

19

k	c	ep	ε	j	t	n	bf	pw	f
343	1	5	L.	3	7	4	18	5	13
344	2	6	J.	5	2	25	9	3	4
345	3	5	M.	6	3	15	29	2	24
346	4	5	Mr.	7	4	30	14	6	9
347	5	5	L.	1	5	22	6	3	1
348	6	6	J.	3	7	11	25	3	20
349	7	5	M.	4	1	26	10	7	5
350	8	5	Mr.	5	2	16	30	6	25
351	9	5	L.	6	3	8	22	3	17
352	10	6	J.	1	5	29	13	1	8
353	11	5	M.	2	6	12	26	7	21
354	12	5	Mr.	3	7	4	18	4	13
355	13	5	L.	4	1	26	10	1	5
356	14	6	J.	6	3	15	29	1	24
357	15	5	M.	7	4	30	14	5	9
358	16	5	Mr.	1	5	22	6	2	1
359	17	5	L.	2	6	12	26	1	21
360	18	6	J.	4	1	26	10	6	5
361	19	5	M.	5	2	16	30	5	25

20

k	c	ep	ε	j	t	n	bf	pw	f
362	1	5	Mr.	6	3	8	22	1	17
363	2	5	L.	7	4	23	7	5	2
364	3	6	J.	2	6	12	26	5	21
365	4	5	M.	3	7	4	18	2	13
366	5	5	Mr.	4	1	19	3	6	28
367	6	5	L.	5	2	9	23	5	18
368	7	6	J.	7	4	30	14	3	9
369	8	5	M.	1	5	20	4	2	29
370	9	5	Mr.	2	6	5	19	6	14
371	10	5	L.	3	7	27	11	3	6
372	11	6	J.	5	2	16	30	3	25
373	12	5	M.	6	3	1	15	7	10
374	13	5	Mr.	7	4	23	7	4	2
375	14	5	L.	1	5	12	26	3	21
376	15	6	J.	3	7	4	18	1	13
377	16	5	M.	4	1	19	3	5	28
378	17	5	Mr.	5	2	9	23	4	18
379	18	5	L.	6	3	1	15	1	10
380	19	6	J.	1	5	20	4	1	29

21

k	c	ep	ε	j	t	n	bf	pw	f
381	1	5	M.	2	6	5	19	4	14
382	2	5	Mr.	3	7	27	11	1	6
383	3	5	L.	4	1	10	24	7	19
384	4	6	J.	6	3	1	15	5	10
385	5	5	M.	7	4	23	7	2	2
386	6	5	Mr.	1	5	13	27	1	22
387	7	5	L.	2	6	28	12	5	7
388	8	6	J.	4	1	17	1	5	26
389	9	5	M.	5	2	9	23	2	18
390	10	5	Mr.	6	3	24	8	6	3
391	11	5	L.	7	4	14	28	5	23
392	12	6	J.	2	6	5	19	3	14
393	13	5	M.	3	7	20	4	7	29
394	14	5	Mr.	4	1	10	24	6	19
395	15	5	L.	5	2	2	16	3	11
396	16	6	J.	7	4	23	7	1	2
397	17	5	M.	1	5	6	20	7	15
398	18	5	Mr.	2	6	28	12	4	7
399	19	5	L.	3	7	18	2	3	27

22

k	c	ep	ε	j	t	n	bf	pw	f
400	1	6	J.	5	2	2	16	7	11
401	2	5	M.	6	3	24	8	4	3
402	3	5	Mr.	7	4	14	28	3	23
403	4	5	L.	1	5	29	13	7	8
404	5	6	J.	3	7	20	4	5	29
405	6	5	M.	4	1	10	24	4	19
406	7	5	Mr.	5	2	2	16	1	11
407	8	5	L.	6	3	15	29	7	24
408	9	6	J.	1	5	6	20	5	15
409	10	5	M.	2	6	28	12	2	7
410	11	5	Mr.	3	7	18	2	1	27
411	12	5	L.	4	1	3	17	5	12
412	13	6	J.	6	3	24	8	3	3
413	14	5	M.	7	4	14	28	2	23
414	15	5	Mr.	1	5	29	13	6	8
415	16	5	L.	2	6	21	5	3	30
416	17	6	J.	4	1	10	24	3	19
417	18	5	M.	5	2	25	9	7	4
418	19	5	Mr.	6	3	15	29	6	24

23

k	c	ep	ε	j	t	n	bf	pw	f
419	1	5	L.	7	4	7	21	2	16
420	2	6	J.	2	6	21	5	7	30
421	3	5	M.	3	7	11	25	6	20
422	4	5	Mr.	4	1	3	17	3	12
423	5	5	L.	5	2	18	2	7	27
424	6	6	J.	7	4	7	21	7	16
425	7	5	M.	1	5	29	13	4	8
426	8	5	Mr.	2	6	19	3	3	28
427	9	5	L.	3	7	4	18	7	13
428	10	6	J.	5	2	25	9	5	4
429	11	5	M.	6	3	15	29	4	24
430	12	5	Mr.	7	4	7	21	1	16
431	13	5	L.	1	5	22	6	5	1
432	14	6	J.	3	7	11	25	5	20
433	15	5	M.	4	1	3	17	2	12
434	16	5	Mr.	5	2	18	2	6	27
435	17	5	L.	6	3	8	22	5	17
436	18	6	J.	1	5	29	13	3	8
437	19	5	M.	2	6	19	3	2	28

24

k	c	ep	∈	j	t	n	bf	pw	f
438	1	5	Mr.	3	7	4	18	5	13
439	2	5	L.	4	1	26	10	2	5
440	3	6	J.	6	3	15	29	2	24
441	4	5	M.	7	4	30	14	6	9
442	5	5	Mr.	1	5	22	6	3	1
443	6	5	L.	2	6	12	26	2	21
444	7	6	J.	4	1	26	10	7	5
445	8	5	M.	5	2	16	30	6	25
446	9	5	Mr.	6	3	8	22	3	17
447	10	5	L.	7	4	23	7	7	2
448	11	6	J.	2	6	12	26	7	21
449	12	5	M.	3	7	4	18	4	13
450	13	5	Mr.	4	1	26	10	1	5
451	14	5	L.	5	2	9	23	7	18
452	15	6	J.	7	4	30	14	5	9
453	16	5	M.	1	5	22	6	2	1
454	17	5	Mr.	2	6	12	26	1	21
455	18	5	L.	3	7	27	11	5	6
456	19	6	J.	5	2	16	30	5	25

25

k	c	ep	∈	j	t	n	bf	pw	f
457	1	5	M.	6	3	8	22	1	17
458	2	5	Mr.	7	4	23	7	5	2
459	3	5	L.	1	5	13	27	4	22
460	4	6	J.	3	7	4	18	2	13
461	5	5	M.	4	1	19	3	6	28
462	6	5	Mr.	5	2	9	23	5	18
463	7	5	L.	6	3	1	15	2	10
464	8	6	J.	1	5	20	4	2	29
465	9	5	M.	2	6	5	19	6	14
466	10	5	Mr.	3	7	27	11	3	6
467	11	5	L.	4	1	17	1	2	26
468	12	6	J.	6	3	1	15	7	10
469	13	5	M.	7	4	23	7	4	2
470	14	5	Mr.	1	5	13	27	3	22
471	15	5	L.	2	6	28	12	7	7
472	16	6	J.	4	1	19	3	5	28
473	17	5	M.	5	2	9	23	4	18
474	18	5	Mr.	6	3	1	15	1	10
475	19	5	L.	7	4	14	28	7	23

26

k	c	ep	∈	j	t	n	bf	pw	f
476	1	6	J.	2	6	5	19	4	14
477	2	5	M.	3	7	27	11	1	6
478	3	5	Mr.	4	1	10	24	7	19
479	4	5	L.	5	2	2	16	4	11
480	5	6	J.	7	4	23	7	2	2
481	6	5	M.	1	5	13	27	1	22
482	7	5	Mr.	2	6	28	12	5	7
483	8	5	L.	3	7	18	2	4	27
484	9	6	J.	5	2	9	23	2	18
485	10	5	M.	6	3	24	8	6	3
486	11	5	Mr.	7	4	14	28	5	23
487	12	5	L.	1	5	6	20	2	15
488	13	6	J.	3	7	20	4	7	29
489	14	5	M.	4	1	10	24	6	19
490	15	5	Mr.	5	2	2	16	3	11
491	16	5	L.	6	3	17	1	7	26
492	17	6	J.	1	5	6	20	7	15
493	18	5	M.	2	6	28	12	4	7
494	19	5	Mr.	3	7	18	2	3	27

27

k	c	ep	∈	j	t	n	bf	pw	f
495	1	5	L.	4	1	3	17	6	12
496	2	6	J.	6	3	24	8	4	3
497	3	5	M.	7	4	14	28	3	23
498	4	5	Mr.	1	5	29	13	7	8
499	5	5	L.	2	6	21	5	4	30
500	6	6	J.	4	1	10	24	4	19
501	7	5	M.	5	2	2	16	1	11
502	8	5	Mr.	6	3	15	29	7	24
503	9	5	L.	7	4	7	21	4	16
504	10	6	J.	2	6	28	12	2	7
50	11	5	M.	3	7	18	2	1	27
506	12	5	Mr.	4	1	3	17	5	12
507	13	5	L.	5	2	25	9	2	4
508	14	6	J.	7	4	14	28	2	23
509	15	5	M.	1	5	29	13	6	8
510	16	5	Mr.	2	6	21	5	3	30
511	17	5	L.	3	7	11	25	2	20
512	18	6	J.	5	2	25	9	7	4
513	19	5	M.	6	3	15	29	6	24

28

k	c	ep	∈	j	t	n	bf	pw	f
514	1	5	Mr.	7	4	7	21	2	16
515	2	5	L.	1	5	22	6	6	1
516	3	6	J.	3	7	11	25	6	20
517	4	5	M.	4	1	3	17	3	12
518	5	5	Mr.	5	2	18	2	7	27
519	6	5	L.	6	3	8	22	6	17
520	7	6	J.	1	5	29	13	4	8
521	8	5	M.	2	6	19	3	3	28
522	9	5	Mr.	3	7	4	18	7	13
523	10	5	L.	4	1	26	10	4	5
524	11	6	J.	6	3	15	29	4	24
525	12	5	M.	7	4	7	21	1	16
526	13	5	Mr.	1	5	22	6	5	1
527	14	5	L.	2	6	12	26	4	21
528	15	6	J.	4	1	3	17	2	12
529	16	5	M.	5	2	18	2	6	27
530	17	5	Mr.	6	3	8	22	5	17
531	18	5	L.	7	4	30	14	2	9
532	19	6	J.	2	6	19	3	2	28

$\overline{m} = 1$

$\overline{yk} = 10$

$\overline{tb} = 15$

$\bar{p} = 14$

mw = pw − 1

Appendix II. Ṭentyon and Concurrentes

When an Alexandrian year A is "intercalary", i. e. when it ends in 6 epagomenal days, then the year A + 1 begins on August 30 of a julian year B. The relationship between Alexandrian and julian intercalations is such, however, that the year B + 1 inserts one day in February which belongs to the year A + 1. Therefore after February the years B + 1 and A + 1 have again the same relation to each other as the years B and A had before the addition of a sixth epagomenal day. In other words: between March 1 and August 29 there exists a fixed relation between

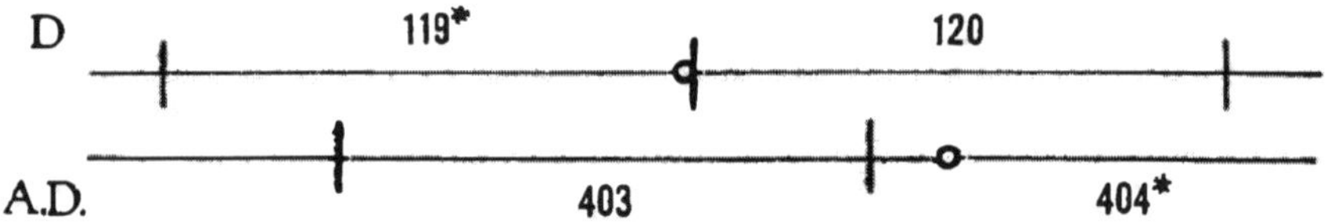

Fig. 1

Alexandrian and julian dates (cf. the subsequent table and Fig. 1). In particular March 24 always corresponds to Alexandrian VII 28 = I 208 ≡ I 5 mod 7[93].

D 117 I, 1 =	A.D. 400* Aug.	29	A.D. 401 Jan. 1 =	D 117 V	6;	March 1 = VII 5
118	401	29	402	118	6	5
119*	402	29	403	119*	6	5
120	403	30	404*	120	5	5
121	404*	29	405	121	6	5
122	405	29	406	122	6	5
123*	406	29	407	123*	6	5
124	407	30	408*	124	5	5

From this relation we can conclude: if I 1 is a Wednesday, i. e. if the ṭentyon $t = 1$, then I 5 as well as VII 28 = March 24 is a Sunday. Since the "feria" of a Sunday is 1, we can say that the feria of March 24 has the same numerical value as the ṭentyon of Thoth 1[94]. The medieval computists called the feria of March 24 the "concurrens"; thus we have shown that its number coincides with the ṭentyon t of the Alexandrian year to which March 24 = VII 28 belongs. We therefore denote in the following also the concurrentes by t, the value of which we can directly

[93] Reference to this relation is found in Abū Šāker, chapter 8 (BM Add 16252 25ᵃ, 20 to 26ᵃ, 2).

[94] Cf. also Schwartz [1904] p. 335 for the role of the feria of Phamenoth (VII) 28.

obtain from the 532-year tables[95]. Note that VII 26 is the earliest possible date for Easter.

In spite of the simplicity of this connection between t and the feria of March 24, one has to face some complications, caused by the unequal lengths of the julian months, if one wishes to determine the weekday of a given date. As an example I shall describe in the following some rules found in an anonymous treatise "de titulis paschalis Aegyptiorum" for the determination of the weekday (pw) of Passover[96].

First step: find the julian date (p) of Passover. In the Alexandrian calendar $p = 10, 29, 18$, etc. for $c = 1, 2, 3$, etc. in each 19-year cycle. In the julian calendar one has to distinguish between March and April. The Alexandrian dates have a simple relation $e + p \equiv 10 \bmod 30$ to the epact (e). Thus one finds

(1)

c	e	p	
1	30	$\overline{\text{VIII}}$ 10 =	Apr. 5
2	11	$\overline{\text{VII}}$ 29	Mar. 25
3	22	$\overline{\text{VIII}}$ 18	Apr. 13
4	3	$\overline{\text{VIII}}$ 7	Apr. 2
5	14	$\overline{\text{VII}}$ 26	Mar. 22

etc. This shows that the day numbers, which decrease always by 11, are in March one unit higher than in the sequence for April. This fact is expressed in our Latin treatise by the rule: we have two "regulares" (r)

(2a)
$$r = 36 \quad \text{for March}$$
$$r = 35 \quad \text{for April}$$

then

(2b)
$$p = r - e.$$

This agrees indeed with (1):

$$\begin{aligned} \text{for } c = 1 \quad & p = 35 - 30 = 5 \text{ April} \\ 2 \quad & \phantom{p = {}} 36 - 11 = 25 \text{ March} \\ 3 \quad & \phantom{p = {}} 35 - 22 = 13 \text{ April} \end{aligned}$$

etc.

[95] Mauro da Leonessa [1943] p. 319 n. 1, incorrectly identifies t with "Regulare Hebdomadae".

[96] Latin text published by Krusch [1937] p. 75 ff. without translation or commentary. Obviously Krusch did not understand the text.

It should be noted that in this context the term "regulares" has nothing to do with other "regulares" known in the medieval computus[97]. Obviously this term has as wide a range as our "parameter" or "coefficient". This will be evident again in the next step which should give us the weekday (pw) of Passover after its date (p) has been found.

As we have seen, the "concurrentes" t give us the feria of March 24. Since

$$\text{Mar. } 0 = \text{Mar. } 24 - 24 \equiv \text{Mar. } 24 - 3 \equiv \text{Mar. } 24 + 4 \mod 7$$

$$\text{Apr. } 0 = \text{Mar. } 24 + 7$$

we can say that

$$\left.\begin{array}{l}\text{weekday of Mar. } 0 \equiv t+4 \\ \text{weekday of Apr. } 0 \equiv t+7\end{array}\right. \mod 7 \tag{3a}$$

hence

$$pw = \begin{cases} p+t+4 & \\ & \mod 7 \\ p+t+7 & \end{cases} \quad \begin{array}{l}\text{if } p \text{ in March} \\ \text{if } p \text{ in April.}\end{array} \tag{3b}$$

The parameters 4 and 7 in (3b) are again called "regulares"[98]. In our example the text takes for the "concurrentes"

$$\begin{array}{cc} c = 1 & t = 4 \\ 2 & 5 \\ 3 & 6 \end{array}$$

etc., and thus finds for the feria of Passover

$$\begin{array}{llll} c = 1 & 5+4+7 = 16 \equiv 2 & & \text{i. e. Monday} \\ 2 & 25+5+4 = 34 \equiv 6 & \mod 7 & \text{Friday} \\ 3 & 13+6+7 = 26 \equiv 5 & & \text{Thursday} \end{array} \tag{4}$$

etc. Since Easter Sunday has the feria 1 (or 8) and since the "luna" of Passover is by definition 14, we have for the "luna" of Easter Sunday in our example:

$$\begin{array}{ll} c = 1 & lu = 14 + (8-2) = 20 \\ 2 & + (8-6) = 16 \\ 3 & + (8-5) = 17. \end{array} \tag{5}$$

These results are, of course, identical with the data directly obtainable from the 532-year table.

---

97 Cf. e. g., Ginzel, Hdb. III, pp. 144—147.

98 The canonical "regulares" give the feria of the first day of the months and are therefore 5 for March and 1 for April.

Christ

See Jesus.

Chronology

See also s. v. Eras (p. 115).

Biographies of Saints or Annals of Kings usually mention, besides regnal years or years of some era, also calendaric parameters, e. g., the number of the epact (*e*), the date of the maṭqeʿe (*m*), or the weekday counted as ṭentyon (*t*) or Day of John (*j*). In the case of Annals such parameters can be tabulated and help to establish correct chronological sequences[1].

A frequently met topic in our sources are biblical chronological lists, covering the 5500 years from Adam to Christ, often expanded into contemporary historical periods[2]. Occasionally the years of the World are subdivided into 532-year cycles[3], up to $13 \cdot 532 = 6916$, leaving 84 years to the second coming of Christ, expected for the year 7000[4].

Many of the 532-year tables[5] contain at the left margin a column headed tārik, i. e. "historical events". There one finds a great variety of entries, usually very unevenly distributed: some biblical dates, some dates from church history (e. g. Councils of Nicaea, Ephesus, etc.) and mainly deaths of kings and patriarchs.

We have even a text[6] where these historical and pseudo-historical data[7] are excerpted for years ranging between 1 and 7225. In each case three calendaric parameters are added: epact (*e*), weekday of I 1 (*tw*), and Evangelist (ε). Also the first years of the consecutive 532-year cycles are indicated (always with $e = 0$, tw = Tuesday, ε = Matthew).

[1] E. g. CSCO 22 to 25 (with several trivial errors, e. g., CSCO 22 p. 3, 33/23 p. 2, 8, a ṭentyon 15 (!) instead of $m = 15$, $t = [3]$, i. e. Friday).

[2] E. g., Berol 84 22ᵇ I, 2—23ᵃ I, 26; Upps 3 63ᵃ, 3—64ᵃ, 16; Vindob 6 2ᵇ, 17—26.

[3] E. g., Princ 5884 7ᵃ I, 4—II, 4. In Ṭānā 34 33ᵇ—92ᵃ and in Vindob 6 4ᵃ the initial years of the 532-year cycles are listed.

[4] Cf. also s. v. Evangelists (p. 128).

[5] Cf. s. v. Calendar, Tables, p. 32.

[6] BN Abb 37, 5ᵃ—8ᵇ.

[7] Conti Rossini [1913] p. 5: "notes, sans intérêt".

Colophons

Cf. s. v. Eras, p. 120 and Calendar, p. 54; Feasts, p. 152.

Computus

See also s. v. Authors.

The term ḥasāb, conventionally translated as "Computus", has a much less well-defined meaning than its medieval counterpart. A great variety of treatises can have ḥasāb in the title, combined with an author's name (e. g., Enoch or Demetrius) or with a topic (winds, kentros, etc.). Furthermore ḥasāb can also denote a treatise of divination or of magic[1]. In short, the term ḥasāb is not characteristic for one single class of texts.

Also with respect to size a ḥasāb is not well circumscribed. It can vary between a few paragraphs or many chapters. Even identical titles (e. g., "Computus of the Stars") may be given to totally different treatises. In the following I give a list of such titles in order to convey to the reader an impression of the variety of topics that can be associated with the concept "Computus". It would be easy to extend this list to a multiple of its length[2].

Computus of Enoch.	Upps 3 60^{a}, 16 to 60^{b}, 9. Multiples of 700 years[3]
Epact.	Upps 3 66^{b}, 13 to 67^{a}, 16. Table of epacts for 19 years
Evangelists.	Upps 3 61^{a}, 7 to 61^{b}, 2. Multiples of 1750 years[4]
Heaven and Earth.	EMML 215 46^{b} II, 22; 47^{a} I, 12, I, 23. This term means the same as "Day of John" (i.e. the parameter j).
kentros.	BM 816 18^{a} II, 17 to 19^{a} I, 4. Length of daylight (12 : 6) Vindob 6 32^{b} II, 16 to 33^{a} I, 16. Length of daylight (15 : 9)
kentros of the Sun.	Berol 84 28^{b} II, 10 to 29^{a} I, 1. Length of daylight (15 : 9)
Law (ḥeg).	Hammerschmidt, Ṭānā 53 (p. 206, III). Calendaric tables
Rotation of the Luminaries.	Enoch 72,1

[1] Strelcyn, Convegno 1959, p. 162 f.; Grébaut [1933] p. 14, 3a; p. 20, 8.

[2] For long lists of titles with ḥasāb, cf. e. g., Ullendorff-Wright, Catal. Cambr. pp. 56—58, or Strelcyn Bibl. Nat., Coll. Griaule, Cat. IV No. 390 ff.

[3] Cf. s. v. Jubilees, p. 172.

[4] Cf. s. v. Evangelists. p. 128.

Sibyl.	Upps 3 60b, 10 to 61a, 7. Cf. s. v. Jubilees (p. 174)
Stars.	Berol 84 32a II, 4 to 33b III, 13. Stars and seasons, etc.
Winds.	BM .816 19a II, 1 to 19b I, 2. Gates, portents of winds.

The whole class of these and similar titles can fittingly be denoted as "Ocean of Computus" (bāḥra ḥasāb)[5]. It would be a major enterprise to establish the relations between all these short calendaric or astronomical excerpts and larger treatises like the "Book of Enoch", the "Guide for the Blind" (marḥa ʿēwur)[6] or the treatises of Abū Shāker or Demetrius.

One group of usually very short treatises has some reasonable internal connections in dealing with the rules for calendaric parameters and dates of feasts. Since all these topics are intimately related to the 532-year tables one may perhaps conjecture that we have here fragments from an "Introduction" to these tables, similar in character to Theon's Introductions to the Handy Tables.

[5] Cf., e. g., Mauro da Leon. [1943] p. 302.

[6] Edited, without translation, by Guidi [1896]; cf. below s. v. Guide for the Blind (p. 165).

Concurrentes

See Calendar, Appendix II (p. 64).

Conversion

The conversion of Ethiopia to Christianity started under King Ezana in the fourth century[1]; Frumentius, the first Bishop of Axum, was consecrated by Athanasius of Alexandria (who died in 373). It is obvious that a considerable time must have elapsed before this process reached a definitive level for the population as a whole. Some traditional dates given in the texts are either obviously wrong (e. g.[2] J 245, i. e. 31 years before Diocletian)[3] or unreliable (e. g.[4] J 430).

[1] Cf. e. g., Ullendorff, Ethiop., p. 54 f., p. 100 f.

[2] Grébaut [1918, 1] p. 325/327; Berol. 84 22b I, 25 to 28; BM 754 4a I, 1 to 3.

[3] This is the correct relation between the eras J and D (cf. s. v. Eras p. 117 Table I and p. 120 Example 2).

[4] CSCO 33 p. 10, 17 f./34 p. 9, 25 to 27.

Dillmann[5] says that the Era of Grace (beginning with J 352 = G 0) "est sine dubio aera Conversionis Aethiopiae". I see nothing that supports this conjecture. The era G is of Alexandrian, not Ethiopian, origin, related to Panodorus' era of the World and to the eras of Diocletian and Incarnation. All epochs for these eras are chosen to be conveniently related to the calendaric cycles. Historical events are likely to be adjusted to the numerological requirements rather than vice versa.

[5] Bodl. Cat., p. 29, colophon to No. 16.

Coptic (or "Egyptian")

See also Greek.

For the transmission of the Alexandrian "computus" (ḥasāb) to Ethiopia one can assume either direct translation from the Greek or a Coptic intermediary (or both). The central piece, the 532-year table (and some of its variants), represents such a consistent unit that a direct copying from the Greek original seems to me the most plausible assumption. A different matter is the problem of transmission of all the explanatory little treatises and rules for the computation of individual parameters (such as epact, ṭentyon, etc.) for which a slow transmission through Coptic (or Arabic versions) seems to be equally possible[1]. In fact this type of auxiliary literature is of very little actual value because it explains hardly anything and the numerical data can all be obtained directly from the 532-year tables.

The Ethiopic texts themselves recognize their dependence on Greek and Coptic sources though one can hardly take such references as historically accurate. For example, in Vindob 6 the 532-year tables are headed by "Of the Copts"[2] in contrast to other texts which credit the "Greeks". A "Computus of the Copts" is mentioned in connection with the date of the Beginning of Fast and of the Solar Epact[3]. Some "Computus of our Fathers, the Copts" describes a lunar calendar of alternating hollow and full months[4] and a shadow table pretends to "measure the hours of Egypt"[5].

[1] Connections with the Coptic curch are, of course, amply attested. The "Book of Hours", e. g., is based on the rules for the hourly offices of the Coptic church (translated to Ethiopic before the 14th century; cf. Cerulli, Lett. p. 27 f.).

[2] Rhodokanakis [1906] p. 82, fol. 4^{a} (zagebṣāweyān).

[3] Berol 84 13^{a} II, 12 (*bf*); 15^{a} II, 14 (*t*).

[4] Berol 84 34^{b} II 7/8; cf. also s. v. Moon, p. 195.

[5] Berol 84 15^{b}.

Appendix. The Coptic "Epact Computus"

In 1942 Georgy Sobhy Bey published (in the Bulletin de la Société d'Archéologie Copte 8 and 9) an English translation and Arabic text of a "ḥasāba alābaqtī", written in the year D 1484 (A. D. 1768). This treatise shows such close parallels to the Ethiopic and thus to the original Alexandrian Easter computus that it can be taken to be typical for sources available in Egypt. Toward the end the text even refers to the Patriarch Demetrius (d. 230), the legendary inventor of the Alexandrian computus[6].

Sobhy Bey had only a very limited understanding of the procedures described in his text and wisely refrained from any commentary. I shall give, in the following, a detailed discussion of all sections which concern our topic, omitting only a few short paragraphs that deal with the relationship between the Coptic and the Arabic calendar. The text itself is not in the best possible condition. The most important topic, the rules for Passover and Easter, are repeated several times with small variations. The arrangement of the sections shows little consistency and many formulations are obscure and could not have been understood had it not been for the experience with the same rules and problems in the Ethiopic treatises.

Needless to say, also the scribe or redactor of the Arabic treatise did not really grasp the reasons for his rules. For example, he explains a number 40 of days as "the number of days in Lent" whereas it is in fact the sum of 30 days (one month) and of 10 days which represent, modulo 30, the distance from maṭqeʿe (m) to Passover (p), $p - m = 190$. This is surely not the only instance where the scribe had not the faintest idea about the origin of his parameters.

The first 5 pages of the treatise are supposedly lost (but the text as we have it begins with the bismillāh). Using the translation of Sobhy Bey I have numbered the sections consecutively from 1 to 30, ignoring the remnants of some other counting. By referring to the sections also with page numbers, I hope to have made them easily identifiable.

Sect. 1 (p. 170/171); cf. also Sect. 6.

Conventional biblical chronology from Adam to Christ (5500), to the Martyrs (additional 276 years, i. e. to the era D), to the "present year" (W 7244).

---

6 Cf. below, p. 92 f.

Computation of the "Norooz"[7] which means here the weekday of New Years Day with the norm of the ṭentyon t. Example for W 7244:

$$7244 \equiv 328 = k \mod 532.$$

The increment in the count of weekdays for k julian years is given by

$$k + [k/4] = 328 + 82 = 410 \equiv 4 \mod 7.$$

Counting from creation[8], i. e. from

$$t_0 = \text{Monday} = 6 \equiv -1 \mod 7$$

one finds for t of W 7244

$$t = t_0 + 4 = 3 \text{ i. e. Friday.}$$

The same result is obtained from the 532-year tables for G = 328.

Sect. 2 (p. 171); cf. also Sect. 8 and Sect. 27.

Find the date of Passover (p). As in the preceding example

$$k = 328 \equiv 5 = c \mod 19$$

and thus[9]

$$(c-1) \cdot 11 = 44 \equiv 14 = e \mod 30,$$

hence[10]

$$p = m + 10 = 30 - e + 10 \equiv 40 - 14 = 26 \mod 30$$

i. e. p = VII 26, since the boundaries of p are VII 25 and VIII 23[11], a norm expressed in the present text as follows:

if $24 < p \leqq 30$	then	p belongs to VII
if $1 \leqq p \leqq 23$	then	p belongs to VIII.

The weekday of VII 1 is given by

$$\text{VII } 1 = \text{I } 1 + 180 \equiv \text{I } 1 + 5 = t + 5 \mod 7$$

hence, with $t = 3$ from Sect. 1:

$$\text{VII } 1 = 3 + 5 \equiv 1 \text{ i. e. Wednesday} \mod 7$$

and therefore

$$p = \text{VII } 26 = \text{VII } 1 + 4 = 1 + 4 = 5 \text{ i. e. Sunday} \mod 7.$$

---

[7] Persian Nauroz for New Year, a term also used by Abū Shāker (Comp. ch. 8, BM Add 16252 27[b], 22; cf. Dillman, Lex. 679).

[8] Cf. s. v. Creation (p. 77).

[9] Cf. s. v. Maṭqe'e and Epact, p. 190 (5a).

[10] Cf. s. v. Feasts p. 131.

[11] Cf. s. v. Feasts p. 133.

Since Passover is seen to fall on a Sunday, Easter will be delayed to $f = p + 7$ = VIII 3 and Palm Sunday, $ho = f - 7$, will coincide with Passover[12].

Sect. 3 to 5 (p. 172/173); cf. also Sect. 10 to 12.

These sections deal with the duration of festival seasons, one called Rifā', the other Fast of the Apostles[13]. The total of these two intervals is always 81 days and it follows from the tabulation in Sect. 27 (p. 192 to 198) that Rifā' $\equiv bf + 31$, $sh \equiv 50 - bf$ mod 30. The Rifā' festival (Carnival) belongs to the months V and VI. In our context these rules are of no interest and will not be discussed in greater detail.

Sect. 6 (p. 173); cf. also Sect. 1

Find the "solar epact" (t). Let Δ be the years from D 1200, i. e. mod 532, from G 60. For G 60 one has $c = 3$, $t_0 = 4$. Then t for the year Δ is given by

$$t = t_0 + \Delta + [\Delta/4] \mod 7$$

where $t = 1$ is Wednesday, not only for I 1 (the "Norooz") but also for VIII 1 = I 1 + 7 · 30.

Sect. 7 (p. 174) concerns the weekday of Christmas.

Sect. 8 (p. 174).

Parallel to Sect. 1, excepting a variant for finding e. Let Δ be the years from D 1200 and reduce $\Delta + 2$ modulo 19. Since $1200 - 2 \equiv 1$ mod 19, the result of this reduction is the cycle number c 1 and therefore

$$e \equiv (c - 1) \cdot 11 \mod 30.$$

Sect. 9 (p. 174).

The epact increases each year by 11 mod 30, excepting when this step-by-step process would result in $e = 29$ which then has to be changed to $e = 30$ (the "saltus lunae").

Sect. 10 to 12 (p. 175/176) concerns the Rifā'; cf. Sect. 3 to 5[14].

Sect. 13 (p. 176/177); cf. Sect. 17.

---

[12] Cf. s. v. Feasts p. 132.

[13] Cf. for the feast day with this name (sh) s. v. Feasts, p. 151.

[14] In Sect. 11 the limits for Passover (cf. above Sect. 2, p. 72, are repeated, but erroneously called limits for Easter.

Weekday of the first day of each Coptic month, but counting from Sunday = 1 (i. e. changing to the norm of the Ethiopic parameter $j = t + 3$)[15]. Hence

for month	I	II	III	IV	etc.
j	$t+3$	$t+5$	$t+7$	$t+2$	

Sect. 14 and 15 (p. 177 to 179) concern the Arabic calendar; cf. Sect. 19.

Sect. 16 (p. 179 to 181).

Multiplication table for the length of the mean synodic month and the lunar year, based on the parameters

$$m = 29;31,50,8,9,24^{d}$$
$$12 \cdot m = 354;22,1,37,52,48^{d}$$

from 1 month to 1174 lunar years[16]. The same parameters are mentioned in a short Amharic section in Vindob 6[17]. According to Bīrūnī these parameters are taken from Ibn Yūnis[18] (around 990). Numbers with so many digits tabulated fall completely outside the calendaric framework of the rest of this treatise. They would only make sense in astronomical lunar tables but then combined with additional tables (e. g., anomaly and latitude). It should be noted that the Arabic lunar calendar is based on the rounded parameters

$$\overline{m} = 29;31,50^{d} \qquad 12\,\overline{m} = 354;22^{d}.$$

Sect. 17 (p. 181). Weekday for any day within a month, based on Sect. 13.

Sect. 18 (p. 182).

Find Easter (f) from the epact (e). Compute

$$m = 30 - e.$$

If $1 \leqq m \leqq 13$, then it belongs to month II; if $14 < m \leqq 30$, to month I (in agreement with the boundaries for the maṭqeʿe I $15 \leqq m \leqq$ II 13)[19]. Then

---

[15] Cf. s. v. Calendar p. 47.

[16] The year A. H. 1174 corresponds to 1760/61, i. e. roughly to the time of writing of our treatise. Computing error: the minutes for the years beginning with 1161 should be 0;1 ^{d} higher.

[17] Vindob 6 3 ^{a}. Same parameters in BM Add 16217 25 ^{a} I, 5/6; EMML 215 4 ^{a} II, 6/7. Cf. also s. v. Arabic p. 18.

[18] Communication by E. S. Kennedy.

[19] Cf. s. v. Feasts, p. 133.

if m is a Saturday	$f \equiv m + 17$	mod 30
a Sunday	$+16$	
etc.		
a Friday	$+11.$	

This is a rule, also found in Ethiopic treatises, for the connection between civil dates and Jewish lunar dates[20].

Sect. 19 (p. 182/183) concerns the Arabic calendar; cf. Sect. 14 and 15.

Sect. 20 (p. 183) concerns the "Greek" (i. e. Syriac) calendar.

Sect. 21 and 22 (p. 184 to 186). Zodiacal signs, lunar mansions.

Sect. 23 (p. 187).

The ṭentyon (t) has a period of $4 \cdot 7 = 28$ years. It is therefore possible to distribute 28 consecutive years into 7 groups of 4 years each, so that each group is associated with the same value of t.

If one chooses the Era of Incarnation (J) as the beginning of the 28-year solar cycle, one has $J\ 1 = W\ 5501 \equiv G\ 13$ mod 28. Therefore all cycle years $k = 13 + n \cdot 28$ (mod 532) in the era G are years with the solar cycle $s = 1$[21]. The only year in this sequence that also coincides with the first year ($c = 1$) in a 19-year cycle is $k = 153$. Inspecting the 532-year table for the years from $k = 153$ to $k = 153 + 28 = 181$ one finds that $t = 1$ is associated with the following four years:

$$\text{(1)} \qquad \begin{array}{rr} k = 153 & s = c = 1 \\ 159 & 7 \\ 164 & 12 \\ 170 & 18, \end{array}$$

and similarly for the other six values of t. By using $k + 76$ we obtain the corresponding year numbers in the Era Diocletian (D) and by taking years $D - 4 = k + 72$, we find for the first year in (1)

$$\text{(2)} \qquad D - 4 = k + 72 = 153 + 72 = 225 \equiv 1 = s \qquad \text{mod } 28.$$

Hence also the subsequent numbers of s will be given by the residues of $D - 4$ mod 28. This is indeed the rule found in our text and explains the corresponding table (p. 187)[22]: If $D - 4 \equiv s$ mod 28, then

---

20 Cf. s. v. Feasts, p. 136.

21 The beginning of a solar cycle at the years $k = 13$ mod 28 is noted in many of our 532-year tables; cf. s. v. Cycles, p. 87.

22 Misprint in col. IV line 2: read 15 instead of 17.

$t =$	1	2	3	4	5	6	7
for $s =$	1	2	3	9	4	5	6
	7	13	8	15	10	11	17
	12	19	14	20	21	16	23
	18	24	25	26	27	22	28

Sect. 24 (p. 187/188).

List of noon shadows for 12 months

month	V	VI	VII	VIII	IX	X	XI	XII	I	II	III	IV	
shadow	9	7	5	3	2	1	1	2	4	6	8	10	feet

obviously badly corrupted. As often in such tables, halves are omitted[23]. Probably the original had the extrema $u = 0$, $U = 9$ and was arranged in six pairs (instead of the correct $1 + 2 \cdot 5 + 1$ pattern), the unfolding of which results in additional errors.

Sect. 25 to 27 (p. 188/189).

Some rules for finding t and e in steps of centuries. Not clear. Followed by the ordinary rules from Sect. 2 for t, e, and p. In the heading of Sect. 27 Abba Demetrius is named as the inventor of these computations, all of which are based on the 19-year cycle.

Sect. 28 (p. 189 to 198).

What follows pretends to be based on an old manuscript found in a desert monastery (Skete), containing "the sayings of Abba Demetrius". A short text describing the events of the Passion is followed by another short paragraph from the Didascalia, warning against the coincidence of Easter with Passover and a celebration before equinox. There is nothing in these concluding sections which would provide us with any specific information about the contents of Demetrius' calendaric work. We are told, however, that in compiling his Easter Computus Abba Demetrius, himself illiterate[24], was inspired by the Holy Spirit. Eventually he communicated his computus to the Patriarchs of Rome, of Antioch, Ephesus, and to his clergy.

There follows a table (in steps of single years) from D = 1401 to 1600. Tabulated are, as function of D, the values of t, e, p, f, $bf + 31$ (i. e.

[23] Cf. s. v. Shadow Tables, p. 213.

[24] Very unlikely in view of the role Demetrius played as Bishop of Alexandria according to Eusebius. Cf. also below p. 93 n. 4.

duration of Rifā'), 50 — *bf* (i. e. duration of Fast of Apostles). All numbers agree with the data given in the Ethiopic 532-year table[25].

Sect. 29 (p. 199).

Length of daylight and night for each month following a linear scheme with $M : m = 15 : 9$ (extrema in IV and X)[26].

Sect. 30 (p. 199). List of Feasts.

---

[25] Excepting several misprints.
[26] Cf. s. v. Length of Daylight, p. 179.

Creation

Biblical tradition places the creation of sun and moon into the fourth day i. e. Wednesday; corresponding statements are often found in our texts[1]. The story was eventually embellished by such details as a creation of the moon in the East at the first hour of night on Tuesday and of the sun, 12 hours later, also in the East, so that the first moon was shining in full brightness[2]. Thus one text[3] tells us that there are three "beginnings" (ṭentāta): "The first beginning, of the day, is Sunday; the second beginning, of the moon, is Tuesday; the third beginning, of the sun, is Wednesday." It is furthermore postulated that this beginning is also the beginning of Maskaram, the first month, and coincides with an equinox[4].

Entirely unrelated to this tradition is the Era of the World, which underlies the whole Easter Computus, i. e. the 532-year tables. According to these tables[5] Maskaram = Thoth 1 of W 1 has the epact 0, thus $m = 30$, which means a new moon ($\overline{m} = 1$) on Maskaram 30. Hence the first day of year 1 (a Tuesday, because of $t = 7$) is also a new moon, not a full moon. Furthermore, the biblical New Year presumably falls into Nisan, whereas Thoth 1 = August 29[6].

---

[1] E. g. Ṭānā 34 29ᵇ (ad year 97); Jubilees II, 8 (Charles, p. 15).
[2] Abū Shāker, Chron. ch. 13: Dillmann, BM. Cat. p. 41/42 (BM Add 16252 44ᵇ); Vat. Cat. p. 388 (Vat 106 13ᵇ II, 14—17).
[3] BM Add 24995 26ᵇ II, 19—24.
[4] Cf. s. v. Seasons, p. 209.
[5] Cf. s. v. Calendar, App. I, p. 59.
[6] Cf. s. v. Eras, p. 118, Table II.

Nevertheless some attempts have been made to connect the Alexandrian calendar[7] with the biblical story. In the Alexandrian calendar the weekday of Thoth 1 (in the norm Sunday = 1) can be found for a year W from[8]

$$j \equiv W + [W/4] + 2 \mod 7.$$

For the Era Diocletian, one has D 1 = W 5777, hence

$$W + [W/4] = 5777 + 1444 = 7221 \equiv 4 \mod 7.$$

But D 1 Thoth 1 was a Friday, hence $j = 4 + 2$.

Instead of this simple explanation of the additive constant 2, our scribes speculated about the biblical story and related the 2 to the two days Sunday and Monday before the creation of the moon on Tuesday[9]. This, perhaps, is the source of a peculiar terminology that refers to these two days as the ʿelata ṭabibān, probably to be rendered as "days of the experts"[10]. Mauro da Leonessa, however, associates three days with this term[11] and explains them as the three days that precede the creation of the luminaries.

---

[7] The calendar "of the Copts" as some texts say (e. g. Ṭānā 34 102^a I, 17; Vat 119 89^a II, 12—14).

[8] Cf. s. v. Ṭentyon and Day of John, p. 222.

[9] Ṭānā 34 102^a I, 26—II, 1; Vat 119 89^b I, 8—11.

[10] Attested in several texts, always with the number 2: BM Add 24995 28^b I, 24; Princ 5884 5^b II, 10; 11^b II, 16; Vat 119 88^b I, 10; etc. Not in Dillmann, Lex.

[11] Mauro da Leon. [1934] p. 312 note 1; I do not know his source.

Cycles

Table of Contents

1. The 19-year Cycle

See also: Calendar (p. 42 and App. I, p. 56); Eras (p. 115).

By the fifth century B. C. Babylonian astronomers had realized that 19 (tropical) years are closely approximated by $19 \cdot 12 + 7 = 235$ mean synodic months. The practical convenience of such a relation suggested to them to regulate their lunar calendar accordingly, i. e. by intercalating one 13th lunar month seven times in a cycle of 19 years. This arrangement agreed well enough with the astronomical facts to dispense with secular corrections, although for theoretical problems (e. g. for the prediction of actual first visibilities of the moon and for eclipses) one made use of still more accurate parameters, among which $29;31,50,8,20^d$ for the length of the mean synodic month is one of the best known parameters of Babylonian and hellenistic astronomy.

The "19-year cycle", which plays such a prominent role in the Easter computus, is a simplified (and much cruder) version of its Babylonian ancestor. It is based on a purely arithmetical pattern that distributes evenly (as far as possible) 235 "lunar months" over 19 civil years. This is done by introducing the "epact" ('abaqtē) which decreases a lunar date by 11^d (mod 30) each calendar year, regardless whether ordinary or leap year, completing the cycle by a final difference of 12^d (in medieval terminology the "saltus lunae")[1] in order to return in cycle year 20 to the date in the first year.

This pattern is attested explicitly in tables[2] that give for 19 years and 12 civil months (and 5 epagomenal days) the dates at which a lunar month begins (cf. Table I). In principle the lunar months alternate between full and hollow (lu, for melu'e, here +, and ḍu, for ḥeḍuḍ, here —). The first month is always full and the dates decrease by 11 days from year to year. In order to maintain this pattern it is necessary, however, to have occasionally at the end of the year (i. e. from XI to XII or from XII or ep to I) two consecutive full months. This happens in seven cases (e. g. from $c = 3$ to 4) when one civil year contains the beginnings of 13 lunar months. Otherwise the alternation between full and hollow is strictly maintained as required for the schematic determination of lunar dates[3].

The following Table II gives, as an example of this procedure, the full moon dates of Passover (*p*) for the years Diocletian 39 to 58 (A. D. 323 to 342). The dates are dates in the Alexandrian calendar, periodically

[1] In BM Add 24995 22^b I, 18—21 the saltus lunae is described as a shortening by one day of the length of the 19th lunar year of the cycle by replacing a "full" month by a hollow one (warḫ ḥeḍuḍa).

[2] EMML 49 85^a—86^a and EMML 215 98^a—99^a (duplicates). These texts abound in scribal errors, the main cause of trouble being the years which contain 13 dates (e. g. for $c = 3$). We have 3 versions of such tables, the initial dates being I 26, I 25, and I 27, respectively. A similar table is given in Mauro da Leonessa, Diss. Cron. p. 86.

[3] Cf. s. v. Feasts, p. 137.

Table I

c	e	I	II	III	IV	V	VI	VII	VIII	IX	X	XI	XII	ep
1	0	26 +	26 −	25 +	25 −	24 +	24 −	23 +	23 −	22 +	22 −	21 +	21 −	
2	11	15 +	15 −	14 +	14 −	13 +	13 −	12 +	12 −	11 +	11 −	10 +	10 −	
3	22	4 +	4 −	3 +	3 −	2 +	2 −	1 +	1/30 − +	30 −	29 +	29 −	28 +	
4	3	23 +	23 −	22 +	22 −	21 +	21 −	20 +	20 −	19 +	19 −	18 +	18 −	
etc.														
18	7	19 +	19 −	18 +	18 −	17 +	17 −	16 +	16 −	15 +	15 −	14 +	14 −	
19	18	8 +	8 −	7 +	7 −	6 +	6 −	5 +	5 −	4 +	4 −	3 +	3 +	3 −

repeated every 19 years. Column Δ shows the length of the intervals that contain 12 or 13 lunar months. Note that 12 lunar months in an ordinary julian year are given 354 days; in a leap year, however, 355 days in order to maintain 11 as the difference in the consecutive dates of p. Only at the end of the cycle is this difference 12, producing one lunar year of 353 days. All intercalary lunar months have 30 days.

It is obvious that this pattern no longer pays any attention either to the mean or to the true synodic month. That the average length of the month implicitly has been made to be $19 \cdot 365;15/235 \approx 29;31,51,\ldots$[d] is of no interest. Nevertheless the resulting pattern agrees fairly well with the average spacing of full- or new-moons, and it is therefore vastly superior to other schemes that are based, e. g., on an 8-year cycle which produces within a few years discrepancies of several days.

Table II

c	D	p	Δ	c	D	p	Δ
1	39*	VIII 10	·	11	49	VIII 20	354 + 30
2	40	VII 29	355	12	50	" 9	354
3	41	VIII 18	354 + 30	13	51*	VII 28	354
4	42	" 7	354	14	52	VIII 17	355 + 30
5	43*	VII 26	354	15	53	" 6	354
6	44	VIII 15	355 + 30	16	54	VII 25	354
7	45	" 4	354	17	55*	VIII 14	354 + 30
8	46	" 23	354 + 30	18	56	" 3	355
9	47*	" 12	354	19	57	" 22	354 + 30
10	48	" 1	355	1	58	" 10	353

The first to make use of the 19-year cycle for the Easter computus seems to have been Anatolius in Alexandria (around 260), later bishop of Laodicea. We know this from a short excerpt from his "Paschal Canon" preserved by Eusebius[4]. Ed. Schwartz assumed that the first cycle covered the years 258 to 276 and attempted to reconstruct its details, partly guided by modern dates[5]. We do not reach safe ground, however, until the Easter letters of Athanasius, which begin in the year 44 of Diocletian (A. D. 328), and with the contemporary Ethiopic 532-year tables[6].

In the Latin church it was not until the fifth or sixth century that the 19-year cycle was adopted. Around 457 the Curia commissioned a cleric, Victorius Aquitanicus, to compile an Easter table. The result was not very successful, probably because Victorius left certain ambiguous Easter dates to the choice of the Pope, which was exactly what one wished to avoid in the controversies with Alexandria.

For us, however, the table of Victorius is of interest because it is based on a 532-year cycle. This cycle begins with the year Victorius postulated for the crucifixion (corresponding to A. D. 28) and ends a century after his time (in A. D. 559). The tables[7] display the following parameters (cf. our Table III): cycle number (k'), for the sake of convenience here taken for the second cycle, in which $k' = 1$ corresponds to D 276 = G 200 $\approx$ A. D. 560; the years in our era; weekday (t') and "luna" (e') of January 1[8]; date[9] and "luna" of Easter Sunday (f and $\bar{f}$). The two last mentioned parameters permit us to determine the date (p') of the "luna 14" because

$$p' = f - (\bar{f} - 14).$$

Table III shows how these data compare with the Alexandrian 532-year cycle.

---

[4] Eusebius EH VII 32, 14 (ed. Bardy II, pp. 225—227).

[5] Schwartz, Ostert., pp. 15—18, repeated by Chaine, Chron. p. 32, who raised, however, serious objections against the validity of Schwartz's reconstruction. It is a misinterpretation of the text when Schwartz ascribes to Anatolius the placing of the vernal point at ♈ 4°; cf. below p. 90.

[6] Cf. s. v. Easter, App. I, p. 98.

[7] Published by Th. Mommsen in Monum. Germ. Hist., Chron. Min. Saec. IV—VII, vol. I (1892) pp. 686—735 and by Krusch [1937] pp. 27—52.

[8] This is an adaptation of the Alexandrian "epact" to the Roman calendar; similarly for t' (cf. s. v. Calendar, App. II, p. 64).

[9] In the original text all dates are given in the Roman calendar, the years (until 559) by consuls. In Table III I have replaced them by the corresponding julian and Alexandrian dates.

Table III

A.D.	k'	f		$\bar{f}$	p'	t'	e'	G	c	p	t	e
560*	1	Mar. 28 =	VIII 2	16	VII 30	5	19	200	10	1	4	9
561	2	Apr. 17	" 22	17	VIII 19	7	30	201	11	20	5	20
562	3	" 9	" 14	20	" 8	1	11	202	12	9	6	1
563	4	Mar. 25	VII 29	16	VII 27	2	22	203*	13	28	7	12
564*	5	Apr. 13	VIII 18	16	VIII 16	3	3	204	14	17	2	23
565	6	" 5	" 10	19	" 5	5	14	205	15	6	3	4
566	7	Mar. 28	" 2	22	VII 24	6	25	206	16	25	4	15
567	8	Apr. 10	" 15	16	VIII 13	7	6	207*	17	14	5	26
568*	9	" 1	" 6	18	" 2	1	17	208	18	3	7	7
569	10	" 21	" 26	19	" 21	3	28	209	19	22	1	18
570	11	" 6	" 11	15	" 10	4	9	210	1	10	2	30
571	12	Mar. 29	" 3	18	VII 29	5		211*	2	29	3	11
572*	13	Apr. 17	" 22	18	VIII 18	6	1	212	3	18	5	22
573	14	" 9	" 14	21	" 7	1	12	213	4	7	6	3
574	15	Mar. 25	VII 29	17	VII 26	2	23	214	5	26	7	14
575	16	Apr. 14	VIII 19	18	VIII 15	3	4	215*	6	15	1	25
576*	17	" 5	" 10	21	" 3	4	16	216	7	4	3	6
577	18	" 18	" 23!	15	" 22	6	27	217	8	23	4	17
578	19	" 10	" 15	18	" 11	7	8	218	9	12	5	28
579	20	" 2	" 7	21	VII 30	1	19	219*	10	1	6	9

The Alexandrian dates of Easter (f) do not require a special column because they agree (with one exception at $k' = 18$) with the dates obtained by Victorius. At $k' = 18$ the Alexandrian table gives VII 30, one week later than the VII 23 of Victorius. The corresponding $\bar{f}$ would be 22 and p' would remain the same. The arbitrary rules for the Easter limits have no effect on the pattern for the phases of the moon.

In comparing the two sides of Table III one must keep in mind that the years of the era G begin 4 months before the years A. D. listed in the same line. Also the counting of the cycles differs since $k' = 1$ begins in $c = 10$. We see, furthermore, that the "saltus lunae" of Victorius is placed after $k' = 16$ which causes the "luna 14" (p') to fall one day earlier than the Alexandrian (p) during the cycle years $c = 7$ to 19. The weekday of Thoth 1 (t) is replaced by the weekday of January 1 ("feria" t' so that 1 = Sunday) and therefore differs numerically from t (Sunday: $t = 5$)[10]. This shows that Victorius was not yet aware of the clever device of using the feria of March 24 (= VII 28), which agrees numeri-

[10] Since there are always 490 days between Thoth 1 of the year G $200 + n$ and A. D. $561 + n$, these years have the same weekday for t and t'. Indeed, for $n = 0$, $t = 4$ = Saturday, $t' = 7$ = Saturday.

cally with t[11]. Finally he tabulates the "luna" of January 1 (column e') which plays the same role as the epact e. Since $e' + p' \equiv 19 \bmod 30$, one can always find from the luna e' the "luna 14" of the Easter full moon.

As a result one can say that the Paschal Table of Victorius is a mediocre adaptation of the Alexandrian 532-year table to the Roman calendar. It is not surprising that 75 years later a consistent version of the Alexandrian 19-year cycle, proposed by Dionysius Exiguus[12], gained decisive influence over the medieval Easter computus.

2. *The Octaeteris*

If one asks oneself how such crude cycles as the "Octaeteris", or the 84-year cycle, could have remained in use for any length of time, one will find two mutually supporting reasons: first, the quality of a cycle was certainly never checked by "observations" and occasional deviations from the actual full moons of one or two days were unavoidable for any cycle and disturbed noone; secondly, the rule that Easter had to be celebrated on a Sunday after the "luna 14" (i. e., p) made it irrelevant what date was provided by a cycle for the full moons (p) unless it happened to coincide with a Sunday. Hence, in the majority of cases, the date of Easter Sunday would not be affected by discrepancies between the arithmetical procedures that provide some date for the "luna 14", not to mention that an error in the prediction of a full moon is much less noticeable than for the first visibility ("luna 1") which can easily be seen to be a full day (or more) early or late.

The Ethiopic computus does not show a trace of any other cycle than the 19-year cycle. It is nevertheless of interest to investigate the working of other cycles because the resulting discrepancies lie at the root of the "Easter controversies" between Alexandria and Rome. It is also of historical interest to see on what primitive, almost "pre-historic", levels such cycles were constructed about one generation after Ptolemy.

An 8-year cycle operates with the round value of $29\frac{1}{2}$ days for the lunar month, the julian year of $365\frac{1}{4}$ days, and with the addition of 3 intercalary months of 30 days each. Then $12 \cdot 29\frac{1}{2} = 354$ and

$$8 \cdot 354 + 90 = 2922^{d} = 8 \cdot 365\tfrac{1}{4}. \tag{1}$$

As an example Table IV shows two consecutive cycles for the dates p of the Easter full moons. Here, and in the following, all dates are dates in the Alexandrian calendar in order to make the arithmetical structure

11 Cf. s. v. Calendar, App. II ("concurrentes") p. 64.

12 Cf. s. v. Easter, App. II, p. 101.

visible and to allow for a direct comparison with data based on the 19-year cycle. The asterisks indicate julian leap years.

Table IV

e′	k′	p′	Δ	k′	p′	Δ
1	1	VIII 18		9	VIII 18	354 + 30
2	2*	″ 7	354	10*	″ 7	354
3	3	VII 25	354	11	VII 25	354
4	4	VIII 14	354 + 30	12	VIII 14	354 + 30
5	5	″ 3	354	13	″ 3	354
6	6*	VII 22	354	14*	VII 22	354
7	7	VIII 10	354 + 30	15	VIII 10	354 + 30
8	8	VII 29	354	16	VII 29	354

The above data are taken from the cycle adopted by Hippolytus in the reign of Severus Alexander, $k' = 1$ corresponding to A. D. 222.

Before comparing this table with the 19-year cycle some remarks about the history of the octaeteris are in order. There exists some evidence for an octaeteris in Alexandria during the third century[13] since the patriarch Dionysius[14], in his paschal letters, is said to have determined the Easter dates from such a cycle[15]. Richard assumed the same for his two predecessors, in particular about Demetrius[16] who then should have transmitted the octaeteris to Hippolytus, Bishop of Rome[17]. Indeed, the Coptic tradition mentions communications between Demetrius and Rome[18] but the procedures ascribed to Demetrius are exclusively based on the 19-year cycle, both according to the Coptic and to the Ethiopic tradition[19].

Attempts to ascribe to the Jews the use of an octaeteris for the determination of the Passover dates[20] ignore the fact that the 19-year

---

13 Any relation to the "octaeteris" of early Greek astronomy seems very unlikely because it was not associated with julian years.

14 Died 264 (Chaine, Chron., p. 251).

15 Eusebius, HE VII, 20 (ed. Bardy II, p. 193).

16 Died about 230.

17 Richard [1974] p. 309.

18 Cf. s. v. Coptic, p. 76.

19 Cf. s. v. Demetrius p. 92.

20 The way in which traces of an octaeteris appear in the Book of Enoch only shows that the author was *not* familiar with the working of this cycle (cf. s. v. Year, p. 232). On Epiphanius cf. Chaine, Chron. p. 27.

cycle ruled for centuries the Babylonian lunar calendar whose first month Nisan has always been also the Passover month for the Jews. On the other hand it is precisely this consistent use of the 19-year cycle by the Jews that could have suggested to the Alexandrian Christians operating with a different cycle in order not to be accused of following Jewish practices in determining the dates of Easter.

3. *Hippolytus*

Eusebius tells us that Hippolytus had written a treatise "On Easter" in which he proposed a 16-year cycle[21]. This story has been fully confirmed by the discovery in 1551 in Rome of a statue of Hippolytus with an inscription that is the equivalent of our Table IV, p. 84.

It is difficult to understand how one could conceive the idea of a 16-year cycle which is nothing but the sequence of two 8-year cycles. Even the most elementary knowledge of lunar motion must have made evident the inadequacy of this "16-year cycle". An estimate of 29;31,50—well known in ancient astronomy—would have shown that 16 years, i. e. $16 \cdot 12+6 = 198$ mean lunar months, contain 5847 days as compared with $2 \cdot 2922 = 5844$ days resulting from (1) p. 83[22]. Richard assumed a modification of the original octaeteris by adding alternatingly 1 day and 2 days to each octaeteris[23], but I see no evidence for a consistent application of such a scheme and, at any rate, it is certainly not part of the cycle of Hippolytus.

The inscription on the famous statue of Hippolytus, now in the Lateran, has been published and discussed many times. The most reliable publications of the text is found in CIG IV, 2 (1859) No. 8613[24]. Scaliger's discussion of the text (in 1595) is reproduced in the edition of 1629 of his "De emendatione temporum", p. 721 ff.

The inscription consists of three parts: A, on the back of the throne, gives the list of the writings of Hippolytus; the calendar tables, B and C, are on the sides. Text B consists of a table of 16 lines; the first column

---

[21] Eusebius, HE VI, 22 (ed. Bardy, II p. 122; cf. also Richard [1966] p. 258).

[22] That such a discrepancy of 3 days in 16 years had been noted is reported by Bar Šinaya (= Elias of Nisibis, around A. D. 1000). Cf. Richard [1966] p. 263/4 from Bar Šinaya; Chabot, p. 120; Delaporte, p. 313.

[23] Richard [1966] p. 267; tabulation for 21 octaeterides from A. D. 221 to 389 in [1974] pp. 334 to 339.

[24] Excluded from IG! Careful presentation also in F. X. Kraus, Real-Encyclopädie d. christlichen Alterthümer, vol. I (1882) pp. 661—663.

gives the dates (in the Roman calendar) of the "luna 14" (ἡ ιδ[25] τοῦ πάσχα) for 16 consecutive years (p'). The next seven columns give the weekdays (pw') of p' for seven consecutive 16-year cycles. The total expansion of the table is therefore $7 \cdot 16 = 112$ years and the factor 7 guarantees the return of the weekdays, but $7 \cdot 8 = 56$ years would have sufficed. Table C contains again 16 lines and 7 columns and gives for each corresponding entry in Table B the date (f') of the respective Easter Sunday. In both tables the years of the first column are the regnal years from 1 to 16 of Severus Alexander (222 to 237).

Both CIG IV and Kraus transpose the Roman dates also to the julian calendar. Much more illuminating, however, is a transformation to the Alexandrian calendar and to the Alexandrian World era (W)[26] because it is these parameters which underlie the original composition of such cycles.

We can now compare the dates obtained by Hippolytus with the results from the Alexandrian 19-year cycle as we know it from the 532-year tables which certainly were in use in Alexandria at least from the fourth century on. It should be underlined that differences between different versions of the computus based on a 19-year cycle would lead to very similar results, since it cannot be the underlying periodic scheme that distinguishes such variants from one another, but only the arbitrary rules about permissible earliest and latest dates for p and f. For the theologians and church politicians this was of primary importance. For the understanding of the arithmetical mechanism, however, such variants are without significance.

Since we are not concerned with the later history of the octaeteris[27] it will suffice to compare the first and the seventh octaeteris (beginning in 222, ending in 278) with the 19-year cycle (cf. Table V; left: octaeteris, right: 19-year cycle). The dates for Passover (p'), for Easter (f'), and for the weekdays (pw') of p' are taken from the inscriptions B and C. Since the weekday of Easter $pw' + (f' - p')$ is found to be always $\equiv 1 \bmod 7$, it is clear that weekdays are counted from $1 =$ Sunday. In the first octaeteris the dates of p' and p agree but 50 years later p' has become senseless. For the dates of Easter one would have obtained full agreement

---

[25] Reversed writing of numerals is not rare in chronological context; cf., e. g. Ševčenko [1967] p. 260, n. 19; p. 263, No. 7, for indictions.

[26] Easter of the year 222 falls into J 214 = W 5714 = G 394 mod 532. This explains the cycle numbers k used in the subsequent tabulations.

[27] For this cf., e. g., the introduction to Jones, Beda.

at the beginning were it not for the different boundary conditions. The table shows that in the Roman norm f' is never permitted to fall on the day after Passover (the "luna 15"), hence $f' - p' \neq 1$, in contrast to the Alexandrian norm. On the other hand VII 22 is permitted for p' but forbidden for p which is restricted to $\geqq$ VII 25.

Table V

k	c′	p′	pw′	f′	f′−p′	p	f	f−p	f′−f
394	1	VIII 18	7	VIII 26	8	VIII 17	VIII 19	2	+7
5	2	″ 7	4	″ 11	4	″ 6	″ 11	5	0
6	3	VII 25	1	″ 2	7	VII 25	″ 2	7	0
7	4	VIII 14	7	″ 22	8	VIII 14	″ 15	1	+7
8	5	″ 3	4	″ 7	4	″ 3	″ 7	4	0
9	6	VII 22	1	VII 29	7	″ 22	″ 27	5	−28
400	7	VIII 10	7	VIII 18	8	″ 10	″ 11	1	+7
1	8	VII 29	4	″ 3	4	VII 24	″ 3	4	0
2	1	VIII 18	3	″ 23	5	VIII 18	″ 23	5	0

k	c′	p′	pw′	f′	f′−p′	p	f	f−p	f′−f
442	1	VIII 18	4	VIII 22	4	VII 26	VIII 1	5	+21
3	2	″ 7	1	″ 14	7	VIII 15	″ 21	6	−7
4	3	VII 25	5	VII 28	3	″ 4	″ 5	1	−7
5	4	VIII 14	4	VIII 18	4	″ 23	″ 25	2	−7
6	5	″ 3	1	″ 10	7	″ 12	″ 17	5	−7
7	6	VII 22	5	VII 25	3	″ 1	″ 2	1	−7
8	7	VIII 10	4	VIII 14	4	″ 20	″ 21	1	−7
9	8	VII 29	1	″ 6	7	″ 9	″ 13	4	−7
450	1	VIII 18	7	″ 26	8	VII 28	″ 5	7	+21

4. *Solar Cycle*

Almost all of our 532-year tables carry at regular 28-year intervals marginal notes da ḍa, i. e. abbreviations of ʿāwda ḍaḥay "solar cycle". Since $28 = 4 \cdot 7$ this cycle repeats the sequence of the Evangelists (ε) as well as the weekday of the first day of the year, i. e. the ṭentyon (t). Obviously the same pair of the parameters ε and t belongs to all cycle years marked with da ḍa.

The solar cycles are not numbered in our tables but they are associated ordinarily[28] with the years

(1 a) $$k \equiv 13 \mod 28$$

i. e., with $k = 13, 41, 69$, etc., for the eras W and G. The choice of $k = 13$ as the initial year is explained by the relation

(1 b) $$\text{G } 13 = \text{J } 365 \equiv \text{J } 1 \mod 28$$

which shows that this solar cycle takes its departure from the era of Incarnation. All these years are years of Mark and $t = 1 =$ Wednesday.

Since

(2) $$153 = 13 + 5 \cdot 28 \equiv 1 \mod 19$$

the year G 153 is not only the first year of a solar cycle but also of a 19-year epact cycle ($c = 1$). No other year in a 532-year cycle has this quality. Obviously for this reason several tables[29] begin with the year G 153. This norm is also known from the Coptic Epact Computus[30].

An exceptional norm is found in Berol 994 (37^b—65^a) where the solar cycles begin at the years $k = 1$, 29, 57, etc. (of the cycle G_3) instead of at $k = 13$, 41, etc. For this norm the solar cycles begin with $\varepsilon =$ Mark, $t = 7 =$ Tuesday.

5. *Great Epact Cycle*

Under this name goes a 95-year period[31] which is apparent in the repetition of Easter dates (modulo 532 years). Strictly speaking this is not a "cycle" because it does not constitute a periodically repeated sequence of numbers. It only means that the four instances of the same Easter date in each 532-year cycle must be found at intervals of 95 years (mod 532). This is a very convenient feature for checking numerical tables and was certainly used for this purpose[32].

---

[28] E. g. in BM 815 3^a—16^b; BM Add 16217 2^a—6^b I; BN Abb 123 184^a—197^b; Ṭānā 34 3^a—16^b; Vindob 6 4^a—17^b.

[29] E. g. BM 816 1^a—5^a; BM Add 16217 2^a—6^b I; Princ 5884 1^a—5^b I, counting G 153 as $k = 1$.

[30] Cf. s. v. Coptic p. 75.

[31] E. g. Berol 84 18^b II, 5—10.

[32] References to this interval are not rare: cf., e. g., EMML 215 59^b II, 2—18; Princ 5884 6^a III, 19—6^b I, 9; Ṭānā 53 107^a I, 4—13.—Modern scholars (e. g. Krusch [1937] p. 14) could have avoided numerical errors had they realized the existence of such a "cycle".

6. Obscure Cycles

Occasionally "cycles" are mentioned for astronomical parameters that cannot be related to familiar periodicities. An example is a "lunar cycle" of "2579 months and (?) 900000 days"[33]. Similarly a cycle of 1250 for the ṭentyon[34] shows no connection with multiples of 7 as one would expect for t. It is of no help that $1250 = 5000/4$.

Astronomically meaningless are two "19-year cycles", one of 6916 days ($= 19 \cdot 364$)[35], the other of 6935 days ($= 19 \cdot 365$)[36], involving the years of Enoch and the Egyptian years. The latter is even extended to a "cycle of Adam" of 194180 days ($= 532 \cdot 365$)[37].

For other "cycles" without astronomical basis, cf. s. v. Jubilees, p. 172.

Appendix. The Cycle of Anatolius

Eusebius in his Ecclesiatical History (VII, 32, 14) gives an excerpt from Anatolius' Easter Canons where we read: "In the first year (of the cycle) the new moon (νουμηνία) of the first month, which is the beginning of the whole 19-year cycle, falls . . . on Phamenoth (VII) 26, according to the Macedonians, however, on the 22nd of the month Dystros . . . (On this date) the sun is not only found inside the first section (of the zodiac) but it had travelled in it already for the fourth day. This section is usually called the first twelfth, and equinox, and beginning of the months, and head of the circle (κύκλου), and start of the course of the planets . . ."

If one wishes to restore on the basis of this passage the 19-year cycle of Anatolius one will note that a lunar date 1 on VII 26 implies for the date 14 of Passover the civil date VIII 9, which is indeed the date of p in the Alexandrian scheme for the cycle year $c = 12$[38]. All subsequent dates will then be obtained by adding 19 (mod 30), with one exception, adding 18, at the "saltus lunae". This discontinuity is located in the Alexandrian pattern between $c = 19$ and 1 which would correspond to $c' = 8/9$ for Anatolius because his $c' = 1$ coincides with Alexandrian $c = 12$. If he furthermore assumed VII 25 and VIII 23 as the boundaries for p his cycle would be identical with the Alexandrian Passover cycle.

---

33 Berol 84 40ᵇ I, 17 to 19.
34 Berol 84 21ᵃ II, 17 f.
35 Vindob 6 23ᵇ I, 7.
36 Berol 84 34ᵇ II, 12 to 39ᵃ III, 28.
37 Dillmann, BM. Catal. p. 4 (No. 5).
38 Cf. above p. 59 Table "XIX".

Unfortunately nothing is sayd in Eusebius' excerpt about the limits for Passover or Easter.

Van de Vyver [1957] suggested a reconstruction of Anatolius' table, assuming that the boundaries for p are the same for Anatolius as in the Alexandrian tables. He furthermore assumed that the saltus lunae was located between $c' = 9$ and 10. If then Anatolius reckoned the epact in the same way as the Alexandrians[39] then one obtains

for $c' = 9$	$e = 1$	$p =$ VIII 11
10	11	VII 29

instead of $e = 0$, $p =$ VIII 10 at $c = 1$ in the Alexandrian scheme. I cannot see it as a great advantage to avoid $e = 0$ (or the equivalent $e = 30$) and to make this the basis for the location of the saltus. We simply do not know where Anatolius put the saltus and how far his pattern differed from the Alexandrian (if at all).

The arithmetical pattern of the 19-year cycle is by itself independent of the date of the vernal equinox. Only if one sets the limits of p into relation with the equinox the question of its date becoms of interest. It is for this reason that the above quoted passage from Eusebius has been much discussed. Schwartz[40] assumed that it implied a location of the vernal point at 4° of Aries, but no such norm is anywhere attested in ancient astronomy. Van de Vyver located the equinox of Anatolius at March 19 (= VII 23)[41] and assumed a later change to March 21 (= VII 25) by the Alexandrians. Again, I know of no parallel of such arbitrary changes in basic astronomical parameters.

The simplest solution seems to me suggested by the text itself. At the beginning Anatolius equates the Alexandrian date VII 26 with the "Macedonian" Dystros 22 and in the rather muddled following discussion[42] he explains that during the "first month" the sun traverses the "first segment" of the zodiac. Such a schematic identification of months and zodiacal signs is quite common in ancient literature and Anatolius could readily say that the Alexandrian scheme for the zodiacal signs preceded by 4 days (or degrees of solar travel) the "Macedonian" norm, i. e. the norm of the Palestinian-Syriac calendar. Hence the whole passage has no astronomical significance whatsoever.

---

39 That is to say that always $e + p \equiv 10 \mod 30$.

40 Schwartz, Ostert. p. 15.

41 Jones (Beda p. 21) assumed a date one day earlier, i. e. March 18 = VII 22.

42 Bardy p. 226 (Euseb. 32, 17).

Dates of Feasts

See Feasts.

Dates of Manuscripts

The majority of the manuscript utilized in the present study belongs to the 17th to 19th centuries. The two oldest manuscripts are Vat 61 (14th cent.)[1] and Upps 3 (15th cent.)[2]. We know practically nothing about the time of the original composition. The modern habit to refer everything that is related to the calendaric computus to "Abū Shāker" (13th cent.)[3] is misleading, since the details of the transmission of the innumerable short treatises scattered throughout our manuscripts have never been investigated. Only for material based on Arabic astronomy and astrology is the original source not in doubt.

Some manuscripts are dated through colophons, which occasionally are quite elaborate[4], but need not concern all parts, in particular not the short chronological and calendaric sections which are inserted where there was space.

Calendaric tables are usually datable through their arithmetical structure, though ordinarily only modulo their characteristic period, e. g., 532 years. Only when these tables also contain the "indictio" does the date become uniquely determined[5].

The 532-year tables frequently carry marginal columns (tārik) which mention events at the respective years and thus provide us with a terminus post quem. But these entries are usually written by several hands and may be much later than the original composition.

Occasionally some short sections within a composite text contain explicit chronological elements. For example, I know of four practically identical copies of a section[6] that concerns relations between 19-year cycles and the "Evangelists" (parameter ε). Only at the very end do these versions differ. The first text says that 99 years elapsed between the death of King Nāod in the year 7000 (once believed to be the year

---

[1] Mauro da Leon. [1943] p. 302.
[2] Löfgren, Kat. p. 31.
[3] Cf. s. v. Authors (p. 22) and p. 15 f.
[4] Cf., e. g., s. v. Feasts, Examples, p. 152.
[5] Cf. s. v. Eras, p. 121.
[6] BM Add 24995 28^b II, 2—22; Princ 5884 11^b III, 9—12^a I, 13; Ṭānā 53 108^a III, 18—108^b I, 12; EMML 2077 3^a I, 1—20.

of the second coming of Christ), two versions put the 28th year of King Fasiladas 152 years after 7000, and the fourth version sets the death of King 'Iasu at 185 years after Nāod[7]. Such dates are embedded like fossils in the conglomerate strata of our manuscripts.

[7] These data agree with Chaine, Chron. pp. 246/7.

Day

See Feasts, Lunar Dates (p. 137); Hours (p. 167); Length of Daylight (p. 179).

Day of John

See Ṭentyon and Day of John.

Degree Division

There is no evidence for a degree division of the circle in the Enochian astronomy. Milik, Enoch p. 14, mentions a 360° division of the horizon (probably by mistake, taken from my analysis of the "gates") and on p. 297 on the basis of an unintelligible fragment that says nothing about degrees.

Demetrius

The twelfth Patriarch of Alexandria, repeatedly credited with the authorship of a "Computus"[1], supposedly written in the year J 206 (A. D. 213/14)[2]; he died A. D. 230[3].

[1] Chaine, Chron. p. 26, n. 3 incorrectly gives bāḥra ḥasāb ("Ocean of Computus") as the title. But BM 815 (25ᵃ II, 10) and its parallels have only ḥasāb (as shown in Wright, Cat. p. 323).

[2] There is trouble with the date: "(Demetrius) consecrated in the 180th year after the Ascension of Christ, he wrote a Computus in the 206th year after the Ascension ('ergata), in the 26th year of his office" (with small variants in BM 815 25ᵃ II, 9—11—incompletely translated by Weld Blundell, Roy. Chron. p. 494, n. 2—BM Add 24995 21ᵃ II, 10—13; EMML 215 2ᵇ II, 23—27; Vindob 6 24ᵃ I, 8—12; cf. also Vat 119 85ᵃ I, 8). If Demetrius died in A. D. 230 $\sim$ J 223 "Ascension" must be an error for "Incarnation", or ergata here means only "appearance".

[3] Chaine, Chron., p. 251.

The tradition about his life and work is extremely unreliable. He is said to have been illiterate[4] but he is also credited with the authorship of an Easter cycle, based on an octaeteris, which he communicated to Rome, thus providing the basis for the cycle of Hippolytus[5]. The Ethiopic treatises, however, which refer to his name, operate consistently with the 19-year cycle and show no trace of an octaeteris.

In the manuscript BN Abb 37 (1^a to 4^b) there is a treatise preserved which represents a fairly consistent introduction to the calendaric concepts and procedures of the 532-year tables[6]. It carries the title ḥasāba zamān "Computus of Time" and is ascribed to Demetrius. In this manuscript there follows a table of historical events[7] and a complete 532-year table, based on the era G. It seems possible that several short calendaric sections which are embedded in a variety of texts are copies of parts of this treatise. Of course there is no real proof of Demetrius' authorship. The problem is further complicated by the fact that a similar treatise, but in details quite different and much shorter, is also ascribed to Demetrius[8].

An Amharic version of the "Ocean of Computus", attributed to Demetrius, is listed by Strelcyn, Catal. Griaule, p. 22 (BN 388 = Griaule 84, 100^a I to 129^b I). A poem in praise of Demetrius as the inventor of the epact is reproduced in Chaine, Chron., p. 70[9].

Appendix. Table of Contents of BN Abb 37, 1^a to 4^b

In the text the beginnings of sections are marked by words written in red. The numbers given to these sections are my addition.

§ 1 Introduction
2 7 basic cycles (cf. § 26)
3 Relations between them

§ 20 *bf*, *mo*, *f* from *m*, etc.
21 Lunar and civil dates
22 Days in 7000 years

[4] Severus al-Muqaffaʿ (according to Renaudot, Hist. Patr. p. 20, p. 24) characterizes him as "vir rusticus, uxoratus, idiota, qui nec legere nec scribere posset". After consecration, however, "tanto studio sacrorum literarum . . . incubuisse, ut doctissimus evaserit". This story is obviously the basis of the Coptic tradition (cf. above p. 76) and of Chaine, Chron. p. 27, end of note 3.

[5] Cf. above s. v. Cycles, p. 84.

[6] For a table of contents, cf. the Appendix (p. 94).

[7] Cf. s. v. Chronology, p. 67.

[8] Extant in a group of parallel texts: BM 815 25^a II, 6; BM Add 24995 21^a II, 6; EMML 215 2^b II, 18; Vindob 6 24^a I, 4.

[9] From Ludolf. Also in Mauro da Leonessa, Diss. Cron. p. 46/47.

4 Multiples of 532 and 7000 years
5 Determination of *e*
6 Explanation of Epact
7 Lunar dates
8 Evangelists (ε)
9 Day of John
10 Length of year
11 Determination of *t*
12 Weekdays from *t*
13 Explanations for *j* and *t*
14 Determination of *m*
15 *e* and *m* for 19 years
16 *m* for consecutive years
17 *yk* and *tb* from *m*
18 *n* from *m*
19 Christian feasts from *n*
23 Equivalences between months
24 Weekdays of beginnings of months
25 Indictio
26 7 basic cycles (cf. § 2)
27 Early and Late Easter dates
28 *f* on VII 29 (Resurrection)
29 Elias
30 "Openings" of Heaven
31 Length of Daylight
32 Public Prayers, dates

On lower margins 2^b to 4^b:

A 7000 years and cycles
B Computus of Bizan, chronologica
C Weeks of Enoch
D Seasons
E *e* and Evangelists (ε)

Diagrams

In several of our manuscripts (all of very late date) one can find crude diagrams, some of them circular, some rectangular, all of them more or less the equivalent of tables. Nowhere do we have real geometrical drawings as they abound in Greek or Arabic mathematical and astronomical treatises. Nevertheless circular arrangement of tabular material is also well known in Greek astronomy, at least since the second century, in the Almagest[1], and later in Byzantine treatises[2]. Rectangular arrangements[3], however, seem to be characteristically Ethiopic.

The detailed interpretation of such diagrams is often very difficult, since a variety of topics (calendaric, zodiacal signs, winds, etc.) are distributed over the whole area, often in a most disorderly fashion, not to mention crowded marginal notes. The following list of diagrams known to me will give a general impression of the material involved. Similar

[1] Cf. Ptolemaeus, Syntaxis Mathematica, ed. Heiberg, vol. I, plate to VI, 11/12, p. 543, 24 f.
[2] Neugebauer, HAMA vol. III, Pl. I.
[3] Cf. Pl. II.

schematic arrangements are also mentioned under "Maps"[4] and under "Orientation"[5].

Berol 994.

68^b:	Square diagram for gates, months, seasons; orientation. Surrounded by circle.

BM 809.

	Pasted on front cover, small circular diagram; better version preserved in EMML 49 and 215 (cf. below).
3^a—4^a:	zodiacal signs, Syriac and Ethiopic months, gates, etc.; lunar phases and lunar illumination; cf. Pl. II.

BM 816.

6^b:	lunar mansions; 4 concentric rings
7^a, 7^b:	zodiacal signs, planetary periods; 8 concentric rings
16^a:	lunar phases; rectangular
16^b:	lunar phases[6], gates; circular [cf. Pl. III]
17^a:	months, winds; rectangular

BM Add 16217.

24^b:	rectangular, with orientation. Horizontal: six gates and months; vertical: hours from 1 to 12.

BM Add 24990.

27^b—28^b:	lunar phases, lunar illuminations, zodiacal signs, Syriac and Ethiopic months; in 3 circular diagrams.

EMML 49 and 215.

27^b/31^b resp.:	small circular diagram, "window", "kēkros" and "kentros", months, directions
28^a/32^a resp.:	months, gates, winds, rivers of Paradise, etc.; rectangular.
Parallel:	BM Add 16217 24^a; Jerus 268 3^a (Littmann [1902]).

Vat. 119 and 128.

55^a/103^a:	winds, evangelists, weekdays; crudely circular (cf. Pl. IV).

---

[4] Below, p. 186.

[5] Below, p. 198.

[6] Ascending and descending lunar "crescents", 15 on each side, are also shown on a Roman monument in Trier (reproduced, e. g., Leroy A. Campbell, Mithraic Iconography, p. 96, Brill 1968; any relationship to Mithra is without a real basis).

Didascalia

Under this name is known an early Christian church order that originated in the third century. The Greek original is lost but versions exist in Syriac[1], Arabic, and Ethiopic. For an English translation of this last version, cf. Harden, Didasc.

In our context only some calendaric rules are of interest. The tendency of the whole work is extremely anti-Jewish and therefore the strict separation between Christian and Jewish festivals is emphasized[2].

The vernal equinox is assumed to fall on VII 25, one day before the earliest date for Easter[3]. This date is said to be VII 22 "according to the Hebrews" and this reckoning should be observed to the 21st (day)[4]. I cannot explain the meaning of VII 22; 21, however, is the upper limit of Easter in the lunar calendar[5]. It is of interest to find this limit, which is characteristic of the Alexandrian computus, as early as the Didascalia.

---

[1] Cf. H. Achelis - J. Flemming, Die syrische Didaskalia, übersetzt und erklärt. TU N.F. 10, 2 (1904). Cf. also Schwartz, Ostert. p. 105.

[2] E. g., Harden, Didasc., p. 130/131; similar, e. g., CSCO 236, p. 90/91.

[3] Cf. s. v. Feasts, p. 134, Table I.

[4] Harden, Didasc., p. 131. Similarly in the maṣḥafa milād (liber nativitatis) of Zar'a Yā'qob (CSCO 221/222 and 235/236), around 1450.

[5] Cf. s. v. Feasts, p. 136. Indeed CSCO 221, p. 115, 11/12 adds expressly "in evening epoch". The translation (CSCO 222 p. 101, 4---8) gives 24 instead of 21 (same error, both in text and translation, CSCO 235, p. 102, 22 and 236 p. 90, 19).

Directions

See Orientation.

Easter

For computational methods concerning Easter, cf. s. v. Feasts (p. 147). Here only some general remarks will be made to clarify the historical background. First: it is a frequently told story that the Council of Nicaea (in 325) laid down rules for the position of the feast within the civil calendar by means of the 19-year cycle. In fact, however, no computational rules concerning Easter are incorporated in the Canones of the Council or in supplementary documents[1].

Secondly: it is widely believed that a determination of the vernal equinox is inherent in the Easter computus. Actually the equinoxes are only indirectly involved through the rules about the civil dates, which should limit the interval permissible for Easter. By forbidding Easter to precede a fixed julian date near the vernal equinox[2], one cut oneself loose in a definitive way from the Jewish pattern in which the 14th of Nisan could oscillate around the vernal equinox in the classical fashion of the Babylonian lunar calendar. If one 14th of Nisan fell later than the equinox and was followed by one preceding it, one "year" (suddenly understood to be determined by the vernal equinox) would contain two Easter celebrations and therefore another year none[3]. No one noted, of course, that the Christian computus must also result in intervals sometimes shorter, sometimes longer, than a julian year. It is clear that political motives lie at the center of interest, not the astronomical determination of the equinoxes[4].

Thirdly: the matter of insistence upon the use of certain cycles is also much more a problem of political prestige than of astronomical efficiency. That empirical data[5] play no role in the arguments between Rome and Alexandria is obvious. For example, the controversy of 455 would have been pointless in relation to the real full moon, which did not occur on the disputed Sunday, April 17, but on Monday, the 18th at 6^h Alexandrian time[6]. The actual lunar motion is much too complicated to be in consistent agreement with any single cycle. Not before the Renaissance, when real astronomy came to life again, was it felt that

---

[1] Cf. Duchesne [1880] p. 23. Dionysius Exiguus in the 6 th century was apparently the first to invoke the authority of Nicaea for the 19-year cycle (cf. Krusch [1937] p. 59; text: p. 63—68) but he cautiously says "non tam peritia saeculari quam sancti spiritus inlustratione sancxerunt" (Krusch, p. 63, 10).

[2] First suggested by Dionysius, Patriarch of Alexandria, about 260. Cf. Richard [1966] p. 268, n. 18.

[3] Cf. Constantin's fulminations (obviously composed by some cleric) in his circular letter at the conclusion of the council. Cf. Duchesne [1880] p. 24.

[4] Duchesne [1880] p. 41: "l'observation de l'équinoxe etait, pour le concil de Nicée, plutot un moyen que le but lui-même".

[5] Krusch, Chron. p, 116: "während man am Himmel Vollmond sah" is an argument that never appears in the texts which operate exlusively with theological sophisms.

[6] Cf. Krusch, Chron. p. 135 with Goldstine, Tables, p. 122.

astronomically significant parameters should be the basis for Easter dates. The consequence of this new approach is the "Gregorian" reform of 1582. For the medieval computus, Alexandrian or Latin, considerations of astronomical accuracy were irrelevant.

As for all schematic astronomical tables there remains the question of the "epoch value", i. e., of the value chosen as starting point for the cycle. It seems to me that an empirical date[7], e. g., of a full moon (even in the ideal case of a lunar eclipse) is out of the question. If the authors of the Alexandrian computus had any knowledge at all of classical astronomy, they would have realized that no single cycle could produce reliable results on the basis of agreement with one arbitrarily picked syzygy. Indeed, Ptolemy had explained at great length how the lunar motion had to be related to distant triples of lunar eclipses in order to determine its parameters[8]. For a cyclic calendar-computation the only astronomically reasonable procedure would have been to relate the cyclic data to the mean syzygies. This would have had the added advantage of an easy solution, since one had excellent tables at one's disposal—e. g., Ptolemy's "Handy Tables" (compiled in the second half of the 2nd century)[9]—that made it very easy to determine the mean syzygies for any given year. Unfortunately there is no way for us to test this hypothesis without any information about the date of compilation of the Alexandrian Easter tables[10].

Appendix I. Athanasius' Easter Dates

The Alexandrian records of the Easter messages of Athanasius cover the period from Diocletian 44 to 89 (A. D. 328 to 373). The calendaric elements which they contain are conveniently tabulated in E. Schwartz's "Ostertafeln" (p. 24 f.). As a result we have at our disposal for 46 years in the fourth century the sequence of all Easter dates (f) and of the age of the moon ($\bar{f}$) on Easter Sunday, of the epact (e) and of a "feria" which

---

[7] An assumption made by Schwartz, Ostert. p. 11.

[8] Cf. Almagest, Book IV.

[9] Theon's version of the "Handy Tables" (compiled near the end of the 4th century) remained in constant use for many centuries.

[10] It is easy to exclude suggested dates. Passover of Diocletian 1, e. g., fell on Pharmouthi (VIII) 10 = 285 Apr. 5, but both mean and true opposition belong to April 7.

is numerically identical with the ṭentyon (t). Implicitly we know also the dates of Passover (p) because $p = 14 + f - \bar{f}$.

These data amply suffice to show that they agree exactly with the Ethiopic 532-year tables. It is, however, a singularly fortunate accident that we have an Ethiopic table[11] that can be dated to the years Diocletian 27 to 85 (A. D. 311 to 369). This date is secured by the fact that not only the cycle numbers k are listed but also the indictions. Consequently the comparison with standard tables does not result in a date modulo 532 years but is unique within 7980 years.

The subsequent excerpts, covering seven years, will suffice to demonstrate the identity of the parameters given by Athanasius and by the 532-year table. The only discrepancy is found in the year 49 where the Athanasian list gives $f = 20$, $\bar{f} = 15$ instead of $f = 27$, $\bar{f} = 21$. But looking at the Passover dates the source of the error is evident: not $f = 20$ but $p = 20$[12]. Furthermore $f = 20$, $p = 20$ would mean $\bar{f} = 14$ (and not 15), a situation excluded because of the coincidence of Easter with Passover. All other dates, excepting trivial scribal errors, agree with the Ethiopic pattern. This agreement for such an early period fully justifies us in considering the Ethiopic 532-year table a faithful replica of the Alexandrian Easter Tables, based on the 19-year cycle from the very beginning, in marked contrast to the instability of the Latin procedures.

The Alexandrian Easter tables have been tabulated for one complete 532-year cycle by E. Schwartz, Ostert., p. 185 to 195. The year numbers which belong to the Easter dates are listed for the following eras:

> "Alexandrinische Osterperiode" (identical with our era G) from 1 to 532 Victorius[13] (from $k = 334$ back to 333); Diocletian (= D) from 77 to 608; Alexandrian World era (= W) from 5853 to 6384; Byzantine World era (from 5869 to 6400); Seleucid era (from 672 to 1203); A. D. 361 to 892.

---

[11] Jerus Arm 3483, 193ᵇ—197ᵃ. The arrangement of the columns differs slightly from the usual type but agrees in all numerical values. The years are called Years of Grace but are in fact years of the Era Diocletian (a frequently found inaccuracy of terminology). The given combination of k and i (e. g. $k = 27$, $i = 14$) can only occur in the cycle D, or in G_3 (cf. s. v. Eras, example on p. 121) and the remaining parameters agree with D, and exclude G_3.

[12] Note that this values is unavoidable because, for the preceding year, we have $p = 14 + 7 - 20 = 1$, for the following year $p = 14 + 12 - 17 = 9$. Hence $1 - 11 \equiv 9 + 11 \equiv 20 \bmod 30$.

[13] Cf. s. v. Cycles, p. 81.

Table I

	D	i	ep	t	j	e	m	$\overline{m}$	yk	$\overline{yk}$	tb	$\overline{tb}$	lf	$\overline{lf}$	p	$\overline{p}$	pw	f	$\overline{f}$
Jerus. Arm. 3483 194b/195a	44	1	6	1	4	25	5	1	14	10	19	15	24	22	15	14	4	19	18
	45	2	5	2	5	6	24*	1	3	10	8	15	16	25	4	14	1	11	21
	46	3	5	3	6	17	13	1	22	10	27	15	29	19	23	14	7	24	15
	47	4	5	4	7	28	2	1	111	10	16	15	21	22	12	14	4	16	18
	48	5	6	6	2	9	21*	1	30*	10	5	15	12	24	1	14	2	7	20
	49	6	5	7	3	20	10	1	19	10	24	15	2*	25	20	14	1	27	21
	50	7	5	1	4	1	27*	1	8	10	13	15	17	21	9	14	5	12	17
Athanasius	44	1		1		25												19	18
	45	2		2		6												11	21
	46	3		3		17												24	15
	47	4		4		28												16	18
	48	5		6		9												7	20
	49	6		7		20												20	15
	50	7		1		1												12	17

The following parameters are tabulated:

$$c \quad s \quad i \quad e \quad t \quad p \quad f \quad \bar{f}.$$

The dates of p and f are given in the julian calendar. For the transformation to the Alexandrian (and Ethiopic) calendar the following relations are conveniently at hand:

(1) March 0 = VII (Phamenoth = Magābit) 4
April 0 = VIII (Pharmuthi = Miyāzyā) 5.

As an example of the usefulness of the 532-year tables we have the dating of four (anonymous) consecutive Easter homilies[14] which provide us with the following Easter dates: April 25, 9, 1, 21, i. e. in the Alexandrian calendar VIII 30, 14, 6, 26, respectively. The 532-year tables[15] show in column f that this sequence occurs once only, for $k = 27$ to 30. The corresponding years are J 379 to 382; hence these homilies concern Easter in the years A. D. 387 to 390[16].

---

14 Floëri-Nautin, Hom. Pasc.

15 Cf. s. v. Calendar, App. I, p. 59.

16 The same result was reached by Floëri-Nautin on ten pages (pp. 18 to 27).

Table II

k: D	G	c	p: A.D.	j.d.	Alex.	Eth.	f: Alex.		Eth.	f̄	t: ecc.	Eth.
229	153	1	513 Apr. 5 =	1908 526 =	$\overline{VIII}$ 10	10	Apr. 7 =	$\overline{VIII}$ 12	12	16	1	1
230	154	2	514 Mar. 25	880	$\overline{VII}$ 29	29	Mar. 30	$\overline{VIII}$ 4	4	19	2	2
231	155	3	515 Apr. 13	1909 264	$\overline{VIII}$ 18	18	Apr. 19	$\overline{VIII}$ 24	24	20	3	3
232	156	4	516 Apr. 2	619	$\overline{VIII}$ 7	7	Apr. 3	$\overline{VIII}$ 8	8	15	5	5
233	157	5	517 Mar. 22	973	$\overline{VII}$ 26	26	Mar. 27[21]	$\overline{VIII}$ 1	30	18	6	6
234	158	6	518 Apr. 10	1910 357	$\overline{VIII}$ 15	15	Apr. 15	$\overline{VIII}$ 20	20	19	7	7
235	159	7	519 Mar. 30	711	$\overline{VIII}$ 4	4	Mar. 31	$\overline{VIII}$ 5	5	15	1	1
236	160	8	520 Apr. 18	1911 096	$\overline{VIII}$ 23	23	Apr. 19	$\overline{VIII}$ 24	24	15	3	3
237	161	9	521 Apr. 7	450	$\overline{VIII}$ 12	12	Apr. 11[22]	$\overline{VIII}$ 16	16	18	4	4
238	162	10	522 Mar. 27	804	$\overline{VIII}$ 1	1	Apr. 3	$\overline{VIII}$ 8	8	21	5	5
239	163	11	523 Apr. 15	1912 189	$\overline{VIII}$ 20	20	Apr. 16	$\overline{VIII}$ 21	21	15	6	6
240	164	12	524 Apr. 4	543	$\overline{VIII}$ 9	9	Apr. 7	$\overline{VIII}$ 12	12	17	1	1
241	165	13	525 Mar. 24	897	$\overline{VII}$ 28	28	Mar. 30	$\overline{VIII}$ 4	4	20	2	2
242	166	14	526 Apr. 12	1913 231	$\overline{VIII}$ 17	17	Apr. 19	$\overline{VIII}$ 24	24	21	3	3
243	167	15	527 Apr. 1	635	$\overline{VIII}$ 6	6	Apr. 4	$\overline{VIII}$ 9	9	17	4	4
244	168	16	528 Mar. 21	990	$\overline{VII}$ 25	25	Mar. 26[23]	$\overline{VII}$ 30	30	19	6	6
245	169	17	529 Apr. 9	1914 374	$\overline{VIII}$ 14	14	Apr. 15	$\overline{VIII}$ 20	20	20	7	7
246	170	18	530 Mar. 29	728	$\overline{VIII}$ 3	3	Mar. 31	$\overline{VIII}$ 5	5	16	1	1
247	171	19	531 Apr. 17	1915 112	$\overline{VIII}$ 22	22	Apr. 20	$\overline{VIII}$ 25	25	17	2	2

Appendix II. The Easter Tables of Dionysius Exiguus

The Easter Tables of Dionysius Exiguus cover the five 19-year cycles from A. D. 532 to 626 and take over an earlier cycle for the years Diocletian 229 to 247 (A. D. 513 to 531) from the tables of Cyrillus[17]. Both tables are arranged in 8 columns, the first of the Dionysian tables denoting the years as "Anni Domini nostri Iesu Christi", an innovation which established the fame of Dionysius as founder of the Christian era (so that Diocletian 248 = A. D. 532).

---

[17] For text and tables, cf. the edition by Krusch [1937] pp. 59 to 74. For notes 21 to 23 cf. pp. 104 and 105.

Table III

I

G		p		f			$\bar{f}$	t	
k	c	A.D.	Alex.		Alex.	Eth.		ccr.	Eth.
172	1	532 Apr. 5 =	VIII 10	Apr. 11 =	VIII 16	16	20	4	4
173	2	533 Mar. 25	VII 29	Mar. 27	VIII 1	1	16	5	5
174	3	534 Apr. 13	VIII 18	Apr. 16	VIII 21	21	17	6	6
175	4	535 Apr. 2	VIII 7	Apr. 8	VIII 13	13	20	7	7
176	5	536 Mar. 22	VII 26	Mar. 23	VII 27	27	15	2	2
177	6	537 Apr. 10	VIII 15	Apr. 12	VIII 17	17	16	3	3
178	7	538 Mar. 30	VIII 4	Apr. 4	VIII 9	9	19	4	4
179	8	539 Apr. 18	VIII 23	Apr. 24	VIII 29	29	20	5	5
180	9	540 Apr. 7	VIII 12	Apr. 8	VIII 13	13	15	7	7
181	10	541 Mar. 27	VIII 1	Mar. 31	VIII 5	5	18	1	1
182	11	542 Apr. 15	VIII 20	Apr. 20	VIII 25	25	19	2	2
183	12	543 Apr. 4	VIII 9	Apr. 5	VIII 10	10	15	3	3
184	13	544 Mar. 24	VII 28	Mar. 27	VIII 1	1	17	5	5
185	14	545 Apr. 12	VIII 17	Apr. 16	VIII 21	21	18	6	6
186	15	546 Apr. 1	VIII 6	Apr. 8	VIII 13	13	21	7	7
187	16	547 Mar. 21	VII 25	Mar. 24	VII 28	28	17	1	1
188	17	548 Apr. 9	VIII 14	Apr. 12	VIII 17	17	17	3	3
189	18	549 Mar. 29	VIII 3	Apr. 4	VIII 9	9	20	4	4
190	19	550 Apr. 17	VIII 22	Apr. 24	VIII 29	29	21	5	5

The following columns tabulate indictio (i), epact (e), concurrentes ($ccr \approx t$)[18], cycle year (c')[19], the date of "luna 14" (i. e. Passover, p),

---

[18] Cf. s. v. Calendar, Appendix II (p. 64).

[19] The counting c' is based on the Byzantine World era and implies that $c' \equiv c - 3 \bmod 19$. In the subsequent tables I replaced c' by the correspond-

Table IV

2

G		p			f			$\bar{f}$	t	
k	c	A.D.		Alix.		Alex.	Eth		cr.	Eth.
191	1	551 Apr. 5	=	VIII 10	Apr. 9 =	VIII 14	14	18	6	6
192	2	552 Mar. 25		VII 29	Mar. 31	VIII 5	5	20	1	1
193	3	553 Apr. 13		VIII 18	Apr. 20	VIII 25	25	21	2	2
194	4	554 Apr. 2		VIII 7	Apr. 5	VIII 10	10	17	3	3
195	5	555 Mar. 22		VII 26	Mar. 28	VIII 2	2	20	4	4
196	6	556 Apr. 10		VIII 15	Apr. 16	VIII 21	21	20	6	6
197	7	557 Mar. 30		VIII 4	Apr. 1	VIII 6	6	16	7	7
198	8	558 Apr. 18		VIII 23	Apr. 21	VIII 26	26	17	1	1
199	9	559 Apr. 7		VIII 12	Apr. 13	VIII 18	18	20	2	2
200	10	560 Mar. 27		VIII 1	Mar. 28	VIII 2	2	15	4	4
201	11	561 Apr. 15		VIII 20	Apr. 17	VIII 22	22	16	5	5
202	12	562 Apr. 4		VIII 9	Apr. 9	VIII 14	14	19	6	6
203	13	563 Mar. 24		VII 28	Mar. 25	VII 29	29	15	7	7
204	14	564 Apr. 12		VIII 17	Apr. 13	VIII 18	18	15	2	2
205	15	565 Apr. 1		VIII 6	Apr. 5	VIII 10	10	18	3	3
206	16	566 Mar. 21		VII 25	Mar. 28	VIII 2	2	21	4	4
207	17	567 Apr. 9		VIII 14	Apr. 10	VIII 15	15	15	5	5
208	18	568 Mar. 29		VIII 3	Apr. 1	VIII 6	6	17	7	7
209	19	569 Apr. 17		VIII 22	Apr. 21	VIII 26	26	18	1	1

the date of Easter (f), and its "luna" ($\bar{f}$). In my subsequent tabulation I omitted column i but added for the sake of comparison with the Ethiopic (i. e. Alexandrian) tables the years of the era G and replaced

ing Alexandrian c. Cf. for c' Rühl, Chron., p. 158; Ginzel, Hdb. III, p. 292; Schwartz, Ostert., p. 21.

Table V

3 G k	c	p A.D.	Alex.	f	Alex.	Eth.	$\bar{f}$	t ccr.	Eth.
210	1	570 Apr. 5 –	VIII 10	Apr. 6 =.	VIII 11	11	15	2	2
211	2	571 Mar. 25	VII 29	Mar. 29	VIII 3	3	18	3	3
212	3	572 Apr. 13	VIII 18	Apr. 17	VIII 22	22	18	5	5
213	4	573 Apr. 2	VIII 7	Apr. 9	VIII 14	14	21	6	6
214	5	574 Mar. 22	VII 26	Mar. 25	VII 29	29	17	7	7
215	6	575 Apr. 10	VIII 15	Apr. 14	VIII 19	19	18	1	1
216	7	576 Mar. 30	VIII 4	Apr. 5	VIII 10	10	20	3	3
217	8	577 Apr. 18	VIII 23	Apr. 25	VIII 30	30	21	4	4
218	9	578 Apr. 7	VIII 12	Apr. 10	VIII 15	15	17	5	5
219	10	579 Mar. 27	VIII 1	Apr. 2	VIII 7	7	20	6	6
220	11	580 Apr. 15	VIII 20	Apr. 21	VIII 26	26	20	1	1
221	12	581 Apr. 4	VIII 9	Apr. 6	VIII 11	11	16	2	2
222	13	582 Mar. 24	VII 28	Mar. 29	VIII 3	3	19	3	3
223	14	583 Apr. 12	VIII 17	Apr. 18	VIII 23	23	20	4	4
224	15	584 Apr. 1	VIII 6	Apr. 2	VIII 7	7	15	6	6
225	16	585 Mar. 21	VII 25	Mar. 25	VII 29	29	18	7	7
226	17	586 Apr. 9	VIII 14	Apr. 14	VIII 19	19	19	1	1
227	18	587 Mar. 29	VIII 3	Mar. 30	VIII 4	4	15	2	2
228	19	588 Apr. 17	VIII 22	Apr. 18	VIII 23	23	15	4	4

all dates of the Roman calendar by the corresponding julian dates and their Alexandrian equivalents. The "Ethiopic" dates for p, f[20], and t (and by implication $\bar{f} = 14 + f - p$) demonstrate that Dionysius simply

[20] Underlined when falling into month VII (magābit), otherwise in VIII.

[21] Luna 18 = 14 + 4 requires March 22 + 4 = March 26 = VII 30 as in the Ethiopic. But Sunday = March 27 (VI Kal. Apr.) = VIII 1. Also Schwartz, Ostert. p. 188 has March 26.

Table VI

4

G		p		f			$\bar{f}$	t	
k	c	A.D.	Alex.		Alex.	Eth.		cer.	Eth.
229	1	589 Apr. 5 =	VIII 10	Apr. 10 =	VIII 15	15	19	5	5
230	2	590 Mar. 25	VII 29	Mar. 26	VII 30	30	15	6	6
231	3.	591 Apr. 13	VIII 18	Apr. 15	VIII 20	20	16	7	7
232	4	592 Apr. 2	VIII 7	Apr. 6	VIII 11	11	18	2	2
233	5	593 Mar. 22	VII 26	Mar. 29	VIII 3	3	21	3	3
234	6	594 Apr. 10	VIII 15	Apr. 11	VIII 16	16	15	4	4
235	7	595 Mar. 30	VIII 4	Apr. 3	VIII 8	8	18	5	5
236	8	596 Apr. 18	VIII 23	Apr. 22	VIII 27	27	18	7	7
237	9	597 Apr. 7	VIII 12	Apr. 14	VIII 19	19	21	1	1
238	10	598 Mar. 27	VIII 1	Mar. 30	VIII 4	4	17	2	2
239	11	599 Apr. 15	VIII 20	Apr. 19	VIII 24	24	18	3	3
240	12	600 Apr. 4	VIII 9	Apr. 10	VIII 15	15	20	5	5
241	13	601 Mar. 24	VII 28	Mar. 26	VII 30	30	16	6	6
242	14	602 Apr. 12	VIII 17	Apr. 15	VIII 20	20	17	7	7
243	15	603 Apr. 1	VIII 6	Apr. 7	VIII 12	12	20	1	1
244	16	604 Mar. 21	VII 25	Mar. 22	VII 26	26	15	3	3
245	17	605 Apr. 9	VIII 14	Apr. 11	VIII 16	16	16	4	4
246	18	606 Mar. 29	VIII 3	Apr. 3	VIII 8	8	19	5	5
247	19	607 Apr. 17	VIII 22	Apr. 23	VIII 28	28	20	6	6

replaced the dates of the Alexandrian table by the corresponding Roman dates. The compilation of his table could not have taken him more than one hour.

[22] Cf. apparatus note f, Krusch [1937] p. 69.

[23] Luna 19 = 14 + 5 requires March 21 + 5 = March 26, which is indeed a Sunday. Thus the text should have VII Kal. Apr. instead of V Kal. Apr. Correct in Schwartz.

Table VII

	G		p		f			$\bar{f}$	t	
5	k	c	A.D.	Alex.		Alex.	Eth.		ccr.	Eth.
	248	1	608 Apr. 5 =	VIII 10	Apr. 7 =	VIII 12	12	16	1	1
	249	2	609 Mar. 25	VII 29	Mar. 30	VIII 4	4	19	2	2
	250	3	610 Apr. 13	VIII 18	Apr. 19	VIII 24	24	20	3	3
	251	4	611 Apr. 2	VIII 7	Apr. 4	VIII 9	9	16	4	4
	252	5	612 Mar. 22	VII 26	Mar. 26	VII 30	30	18	6	6
	253	6	613 Apr. 10	VIII 15	Apr. 15	VIII 20	20	19	7	7
	254	7	614 Mar. 30	VIII 4	Mar. 31	VIII 5	5	15	1	1
	255	8	615 Apr. 18	VIII 23	Apr. 20	VIII 25	25	16	2	2
	256	9	616 Apr. 7	VIII 12	Apr. 11	VIII 16	16	18	4	4
	257	10	617 Mar. 27	VIII 1	Apr. 3	VIII 8	8	21	5	5
	258	11	618 Apr. 15	VIII 20	Apr. 16	VIII 21	21	15	6	6
	259	12	619 Apr. 4	VIII 9	Apr. 8	VIII 13	13	18	7	7
	260	13	620 Mar. 24	VII 28	Mar. 30	VIII 4	4	20	2	2
	261	14	621 Apr. 12	VIII 17	Apr. 19	VIII 24	24	21	3	3
	262	15	622 Apr. 1	VIII 6	Apr. 4	VIII 9	9	17	4	4
	263	16	623 Mar. 21	VII 25	Mar. 27	VIII 1	1	20	5	5
	264	17	624 Apr. 9	VIII 14	Apr. 15	VIII 20	20	20	7	7
	265	18	625 Mar. 29	VIII 3	Mar. 31	VIII 5	5	16	1	1
	266	19	626 Apr. 17	VIII 22	Apr. 20	VIII 25	25	17	2	2

Eclipses

Solar Eclipses

1. Bibl. Nat. No. 126 (Catal. Zotenberg, p. 158, 9. d.) "grande eclipse du soleil qui eut lieu en l'an 958 (?) [text: 905] des martyrs, sous le règne de Ṣalaḥ Ayyoub . . . et sous le patriarcat de Cyrille". According to Chaine, Chron. p. 244 and 253:

> Ayoub Nedim ed Din el Melek el Sāleḥ: 1239—1249 (an Ayubid ruler)
> Cyrille (III): 1235 June 17 to 1243 March 10 or 12 (Patriarch of Alexandria).

During the overlapping years, 1239 to 1243, however, no solar eclipse was total in Ethiopia[1], but in 1241 Oct. 6 one—total in the Persian Gulf—may have been noticed, hence the emendation D 958 (= J 1234 = A. D. 1241).

2. In the Liber Axumae we have a long list of dates of important events[2], reckoned in the era Diocletian[3] though called "years of Grace". One of the entries says "in the year of Grace 180 a solar eclipse". Now D 180 $\equiv$ D 1244 mod 532, and D 1244 = J 1520 = A. D. 1527/8. Indeed there occurred a solar eclipse, total in Ethiopia, in A. D. 1528 May 18 = J 1520 Genbot (IX) 23[4].

3. The historical column of the 532-year table in BM 815[5] mentions for the year $k = 304$, i. e. for G_3 304 = J 1720 = A. D. 1727/8 a "solar eclipse". Indeed the eclipse of 1727 Sept. 15 was total for Ethiopia[6].

4. In the Chronicle of John, Bishop of Nikiu[7], a solar eclipse is mentioned[8] which occurred in the reign of Mauritius (582—602) and was total in Antioch in the fifth hour of the day, at a time when this city

---

[1] The nearest date would be A. D. 1235 Aug. 15 (= J 1227 = D 951).

[2] CSCO 54, p. 67 f./58 pp. 81—83, etc.

[3] This is evident, e. g., from known regnal years.

[4] Chaine's table (Chron. p. 272) gives incorrectly the year J 1521 = D 1245.

[5] Fol. 10ᵇ.

[6] Oppolzer, Canon No. 6979.

[7] In the delta, not far from Cairo. Cf. above s. v. Authors p. 25 n. 28.

[8] Text: Zotenberg [1883] p. 302, trsl. p. 536 (ch. 101). Also in the table of contents, Zotenberg, Cat. p. 237 (BN 146) = [1883] p. 144, ch. 100. English trsl. Charles, Chronicle, p. 163.

experienced a severe earthquake[9]. The given data fit excellently the eclipse of 601 March 10, but not the additional statement that the event took place at the end of the twelfth 532-year cycle which would be the year J 884 = A. D. 891/2.

This second date, however, is also connected with a solar eclipse, total at noon in Antioch, in the year 891 August 8[10]. Obviously a later scribe or translator substituted information from a list of eclipses, forgetting the incompatibility of its date with the reign of Mauritius or the lifetime of John of Nikiu.

5. Chaine, Chron., p. 271 f., gives a list of solar eclipses, visible in Egypt and Ethiopia, between A. D. 20 and 1882 (excerpted from Oppolzer, Canon). The years are listed in the eras A. D., J, and D, the days in the Ethiopic and Coptic calendar. These transformations are, however, not reliable[11].

Lunar Eclipses

I know only of references to one lunar eclipse (ḍalma warḫ) from the margin of 532-year tables[12]. The date is $k = 197$ in the cycle G_3, i. e. A. D. 1620/21. Indeed a total lunar eclipse (magn. 19.2) took place 1620 Dec. 9, though only its beginning could have been seen in Ethiopia[13].

---

[9] Grumel, Chron., p. 479 lists an earthquake in 601 Apr. 2 in Mesopotamia.

[10] In local time Baghdad, the hours of these two eclipses are 11;10 and 13;7, respectively; cf. Goldstine, New and Full Moons 1001 B. C. to A. D. 1651, p. 134 and p. 158.

[11] Cf. the preceding note 4.

[12] Berol 994 47[b], BM 815 8[a] $\sim$ BM 827 109[a]. These are the same texts which report a Nova for $k = 195$ (cf. below s. v. Nova, p. 198).

[13] Oppolzer, Canon No. 4367; Schroeter, Kanon No. 589.

Egypt

See Coptic.

Elias

See Feasts (p. 151).

Enoch

Qui a fait croire aux gens simples que ces livres sont d'Enoch?
Athanasius (CSCO 151, p. 38, 4)

A short section in Genesis[1] is the basis for the "Book of Enoch" which indulges in an elaborate description of the patriarch's journey into heaven under the guidance of an angel. The biblical text has no astronomical implications and even the astronomical chapters (72 to 82) in the extant version of the Book of Enoch clearly originated as an independent treatise which was only secondarily incorporated into the Enoch legend[2]. The title of this section is "Book on the Changes of the Luminaries" (miṭata berhānāt) and is frequently quoted in our treatises[3] but never as part of the "Book of Enoch".

In contrast to the Ethiopic calendaric treatises that concentrate on the dates of the moveable festivals, Christian as well as Jewish, the astronomical sections of the Book of Enoch have a mainly didactic tendency in dealing with general astronomical matters which seemed of interest to the leaders of a religious sect. The level of these discussions is extremely elementary in sharp contrast to contemporary Babylonian or hellenistic astronomy.

The main topics of the astronomy of "Enoch" are easily enumerated: the risings and settings of sun and moon through the "gates" on the horizon and the linear variation of the illumination of the moon during a schematic lunar month[4]; another linear scheme for the changing length of daylight with the extrema $M : m = 2 : 1$ and $M + m = 18$ parts[5]; a "year" of 364 days, divided into four seasons of $30 + 30 + 30 + 1 = 13 \cdot 7$ days each. This departure from the traditional Near Eastern lunar calendar towards a (only slightly modified) Egyptian-hellenistic pattern is perhaps the most remarkable feature of this heretical calendar[6].

---

[1] Genesis 5, 18—24.

[2] Ethiopic sources (e. g. BN Abb 37 3^b, lower marg. 4) assign the life of Enoch to the years $6 \cdot 700 + 100$ (= 4300) before the birth of Christ.

[3] E. g. Berol 84 41^b III, 12/13; Berol 994 68^a II, 4/5, 68^b diagram; BM Add 16217 14^a I, 3/4; EMML 215 49^a II, 18/19; Princ 5884 14^b III, 7/8. In relation to eclipses: BM 816 18^a II, 5/6.

[4] Cf. s. v. Moon (p. 195).

[5] Cf. s. v. Length of Daylight (p. 180).

[6] I do not understand how Barthélemy [1952] p. 201, n. 4 could say: "On sait que tout le monde hellénestique utilisa le mois lunaire avec mois

The Book of Enoch existed in late pre-Christian times in an Aramaic version, as we now know through the discoveries in Qumrān[7]. Nevertheless the Ethiopic version is the only complete version[8]. Almost nothing is known about the time and the circumstances of its transmission into Ethiopia[9]. The flagrant discrepancies with conditions in Ethiopia (e. g. the extrema 2 : 1 for the length of daylight) and the incompatibility of its calendaric system with the established Alexandrian calendar did not detract from the esteem in which the Book of Enoch was held. References to it abound in Ethiopic texts[10] and Zar'a Yā'qob (around A. D. 1450) could say[11]: "No man, may he come from East or West, from North or South, from all the four corners of the world, can compute the time of Fast, of Easter and its feasts, or of the sun or the moon, or the stars, without Enoch." And two medieval versions still exist, translated from the Greek to Church-Slavonic[12].

The arithmetical scheme that relates full and hollow months alternatingly to the "gates" is not found in the Book of Enoch as we have it. But the short treatises which display the numerical details[13] are based on the concept of gates known from the Book of Enoch, so that a common origin seems likely. Explicit references to Enoch's authorship can therefore be accepted as correct. This supports the conclusion that we do not have the complete version of the astronomical treatise that eventually was absorbed into the "Book of Enoch".

---

intercalaire periodique jusqu'à l'introduction du calendrier julien". This drastic misconception then leads to the idea (p. 201 ff.) that the Jewish lunar calendar is a compromise with hellenistic customs.

7 Milik, Enoch, p. 6, p. 22, etc.—A good example for the influence of the "Enoch" calendar within the world of Qumrān is found in an enumeration of Psalms composed by David. I quote from a translation by Dupont-Sommer (CRAIBL 1977, p. 547): 364 Psalms for the daily sacrifices, 52 for the Sabbaths, 30 (for the days of the months), 4 (for the days added at the end of the 4 seasons—not for "personnes frappées [par des esprits mauvais]"!), making a total of 450 (not 4050) Psalms.

8 Text: Dillmann, Liber Henoch, Aethiopice, 1851; Flemming, Henoch (1902).

9 Translation: Charles, Enoch (1893); Flemming-Radermacher, Henoch (1901). Cerulli, Lett. p. 20 f., places the time of translation into the period from the 4th to the 7th centuries.

10 Cf. the long list of references assembled by Milik, Enoch, p. 86 f. This list does not include the (mostly unpublished) calendaric material.

11 CSCO 235, p. 99, 10—14 and 236, p. 87, 17—21 (with some inaccuracies).

12 Cf. below, Appendix p. 111.

13 Cf. s. v. Gates (p. 160) Table I.

From the viewpoint of the history of astronomy Enoch's revelations are not very interesting. Preoccupation with risings and settings is a characteristic feature of all primitive astronomy. No attempt is made to describe solar or lunar motion; no mention is ever made of the planets and also the stars are not more than decorations of the celestial palace.

The most characteristic element of Enoch's astronomy is the year of 364 days and its division into four seasons of 13 weeks each. Obviously it is the fixed coordination of days of the week and calendar dates which recommended this norm to its inventors. Since a year of 364 days will traverse in three centuries all seasons of the solar year, modern scholars have tried to discover evidence for an intercalation scheme that would counteract this consequence. Such hypothetical arrangements, however, disregard the fact that all our texts consider a year of 364 days as typical for the "year" of Enoch. It may suffice to quote the statement that 7000 years contain 2 548 000 days (i. e. $7000 \cdot 364$), in contrast to the years "with us" which total 2 556 750 ($= 7000 \cdot 365\frac{1}{4}$) days[14]. Such a comparison would be meaningless if any intercalation would ever interfere with the fixed length of the year of Enoch.

Our texts[15] name Enoch also as the author of another work that is fully preserved only in Ethiopic, the "Book of Jubilees" (accepting a Hebrew title[16]: the Ethiopic title is "Book of Partition [kufalē] of the Days"). It was written in the second century B. C.[17] and contains very little of astronomical or calendaric interest.

Appendix. The Slavonic Enoch

A short treatise with the title "The Secrets of Enoch" has survived in two Church Slavonic versions of unequal length[17a]. The original, now lost, was written in Greek, probably composed around 900 by a Byzantine monk[17b]. As Vaillant has shown the shorter Slavonic translation represents the original version which was amplified by a second redactor some time between the 13th and 16th centuries, or later[17c].

---

[14] E. g., BM Add 16217 17[a] upper margin I/II.

[15] EMML 215 48[b] I, 10—20; Marḥa 'E 373 II, 5; 374 II, 26; 382 II, 26—27; 383 I, 17.

[16] Cf. Charles, Jubilees, Introd. § 2.

[17] Charles, l. c., § 17.

[17a] Translations: Charles-Morfill, Enoch (1896); Bonwetsch [1896] and [1922]; Vaillant, Hénoch (1952).

[17b] Cf. Milik, Enoch, p. 112.

[17c] Maunder [1918], Fotheringham [1922]; Milik, Enoch, p. 108.

The contents of the "Secrets" can be described as a paraphrase of the original "Books of Enoch". In the present context only the astronomical passages need to be considered. By distinguishing between the shorter and the longer versions one realizes the trend to modernization in spite of the fact that elements from the original Enochian tradition, e. g., the 364 day year, are carefully preserved.

Short version[18]:

XI (V. p. 11/13).
Brightness of the sun in relation to the moon (cf. s. v. Moon, p. 196).

XIII (V. p. 13).
Six gates each, east and west. Each gate is 30 stades wide (thus the solar travel during one year, traversing twice all gates, amounts to 360 stades, i. e. degrees; cf. s. v. Year, p. 231).
Risings of the sun in 6 gates: $8 \cdot 35 + 2 \cdot 42 = 364$ days, i. e. year of Enoch; cf. s. v. Gates, p. 157.

XVI (V. p. 15).
Twelve gates for the moon, actually the length of the 12 months of the julian calendar (with some errors, caused by the scheme in ch. XIII; cf. s. v. Gates, p. 157).

Longer version:

XIII (V. p. 91).
Six gates east and west. Each gate is $61\frac{1}{4}$ stades wide. Perhaps to be emended to 5 gates of 61 stades and one of $60\frac{1}{4}$ stades, thus a total of $365\frac{1}{4}$ stades, corresponding to a solar motion of 1 stade per day.

XIV (V. p. 91/93).
Year of $365\frac{1}{4}$ days.

Length of night $1 + 7 + 1 = 9$ "great hours", i. e. nychthemeron of 18 parts (cf. s. v. Length of Daylight, p. 180).

XV[19].
28-year solar cycle.

---

[18] The Roman numerals give the chapters in Bonwetsch [1922] whereas V. refers to Valiant, Hénoch.

[19] Not in Vaillant; cf. Bonwetsch [1922] p. 14.

XVI (V. p. 93).

Year of 365¼ days, lunar year of 354 days, epact of 11 days, 4-year intercalation cycle.

532-year luni-solar cycle; 19 years of 19 lunar years and 7 intercalary months[20].

XXX (V. p. 99/101).

List of the planets in a peculiar order (cf. s. v. Planets, p. 201). Zodiacal signs, brontologia (?), nativities, horologia.

XLVIII (V. p. 109).

182 thrones for "small" and "large" days each, i. e. winter + summer = 364 days.

Solstices: Thebet (IV) 17 and Thammuz (X) 17; same dates in Vat. 54, 88ª 9 to 89ª 19, a text written under Arabic influence[21].

---

[20] This is the obvious meaning of the "seven" which caused trouble both to Bonwetsch (p. 15, 19) and to Vaillant (p. 92, note 7).

[21] Parallel: Ambros B 22 182ª 2 to 182ᵇ 11; 1ª to 2ª 2 (incomplete).

Epact

See Maṭqeʿe and Epact (p. 187).

Epagomenal Days

The term paguemēn is obviously derived from the Greek ἐπαγόμεναι (ἡμέραι) which refers to 5 or, every fourth year, to 6 days at the end of the calendar year[1]. In general this position remains the same also in the Ethiopic calendar; in the shadow tables, e. g., the epagomenal days are counted with month XII[2]. It is a misnomer of modern scholarship to consider the epagomenal days as a "13th month".

In the calendaric tables, however, the epagomenal days are associated with the subsequent civil year. This is evident, e. g., from the following excerpt from the 532-year tables

---

[1] A short note in Vindob 6 18ª (parallels: Princ 5884 7ª III, 18—20; Ṭānā 53 107ᵇ I, 24—26) gives the equivalent names in "Persian, Coptic, and Roman" which look, however, more like arbitrary inventions.

[2] Cf. below, s. v. Shadow Tables, p. 211 n. 6.

k	ep	tw	ε	j	t
1	5	Tuesday	Matth.	3	7
2	5	Wednesday	Mar.	4	1
3	5	Thursday	Luke	5	2
4	6	Saturday	John	7	4
5	5	Sunday	Matth.	1	5

Since $365 \equiv 1 \bmod 7$ consecutive weekdays indicate ordinary years. The gap of 2 days, however, shows that 6 intercalary days are located between the years "Luke" and "John" and the column "ep" connects them with "John"[3], i. e. with the immediately following 12 months. This order is also explicitly attested in several of our treatises[4] which tell us "Because these (the epagomenal days) were first created with the 12 months, until the present time, they precede (yeqademā) the start of each year". Hence one could simply say that the "epagomenal days" precede the calendar year. Cf. also Fig. 2 where the small circles indicate the position of Maskaram 1, the civil New Years Day[5].

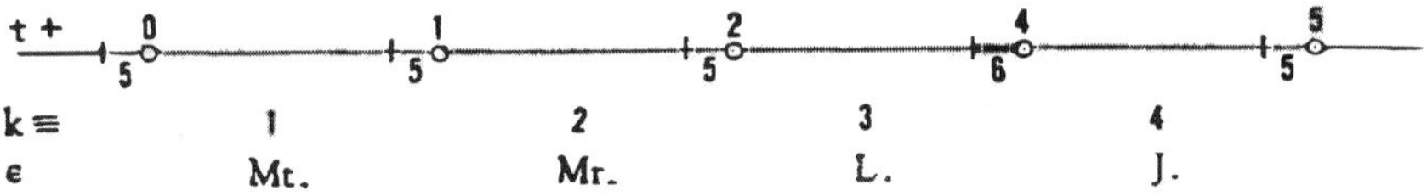

Fig. 2

The year of Enoch has no epagomenal days in the proper sense of this term. Instead every third month is given 31 days in order to produce a year of four seasons of 91 days each. Nevertheless the Ethiopic authors try, as usual, to compress contradictory traditions into one system. This results in very unclear discussions in which somehow the change of the

[3] Consequently one speaks of "intercalary days of John" (ʿelat sagra yoḥanes), e. g., Upps 3 61 ᵇ, 10. Ordinary years are ʿāmata manfaq, "intermediate years" (e. g. BM Add 16252 137 ᵃ, 9—13).

[4] Berol 84 3 ᵃ I, 6—9; Upps 3 69 ᵇ, 14—70 ᵃ, 1; Vat 1 208 ᵇ II, 28—32. Similarly in the computational rules, e. g. Vat 119 89 ᵇ II, 1—4; if $W \equiv 0 \bmod 4$, then $\varepsilon =$ John and 6 epagomenal days.

[5] In the years of John ($k \equiv 0 \bmod 4$) this day corresponds to the julian August 30, in all other years to August 29; cf. also below s. v. Leap years, p. 179.

length of daylight and night is made responsible for the addition of one day to bring the 364-day year of Enoch at least up to 365 days[6].

As Henning has shown[7] the influence of the Book of Enoch has reached even Mani (in the early 3rd century A. D.). In the fragments of Manichaean literature found in Turfan occur statements[8] in which the epact 364 — 354 = 10 days of the Enoch year is compared with the difference 366 — 354 = 12 for an Alexandrian leap year. This is basically the same approach as in the Ethiopic version.

---

[6] Berol 84 41^{a} I, 26—41^{b} III, 14; Princ 5884 14^{a} I, 1—14^{b} III, 10 is a parallel for the section beginning at 41^{a} III, 1.

[7] W. Henning, Ein manichäisches Henochbuch. SB d. Preußischen Akad. d. Wiss., Philos.-hist. Kl., 1934, pp. 27—35.

[8] Henning, l. c., p. 34 f.

Equinoxes

See Seasons (p. 204).

Eras

Table of Contents

1. *The Ethiopic Eras*

The Ethiopic technical term for "era" is ḥasāb, literally "computus"[1]. We read, e. g., in Abū Shāker's chronological treatise, Chapter 10: "There are 3 eras (ḥasābat), the first is the era of the Flood, the second the era Nabonassar, the third the era Philip."[2] Similarly, Chapter 9 discusses the combination of the intervals between different eras[3].

---

[1] This technical usage is not recognized in our dictionaries.

[2] BM Add 16252 31^{b}, 8 to 10.

[3] BM Add 16252 30^{a}, 12 to 31^{a}, 7.

In the present context, however, neither the well known historical eras nor the Biblical chronology need be discussed. For the understanding of Ethiopic calendaric treatises and tables (and of colophons) it will suffice to give an accurate account of the interrelations between the four major eras which occur in one form or another in almost every text. The transformation into the conventional julian era of modern historiography is in all cases a trivial operation.

The eras on which Ethiopic chronological data rest are all of Alexandrian origin and ultimately based on the Era Diocletian. Nevertheless Ethiopic texts only rarely associate the name of Diocletian with the years of his reign[4]. Instead, they speak about the "years of the Martyrs" (ʿāmata samāʿetāt) or of the "years of Grace" (ʿāmata meḥrat)[5], although this latter term denotes also the years of an era which begins 76 years after the Diocletian era. This ambiguity of terminology has often marred modern interpretations. We therefore shall use the letter D exclusively for the era Diocletian, the letter G for the era defined by

(1) $$G\ 1 = D\ 77,$$

regardless of the terminology used in the texts. Both for D and G the years 3, 7, 11, etc. are leap years and both eras are subdivided into cycles of 532 years. If necessary we can distinguish these cycles as D_1, D_2, D_3 and G_1, G_2, G_3 respectively. Hence G_1 covers the years D 77 to 608 and similarly for the other cycles. The majority of all preserved texts belongs to the cycle G_3.

Two more eras are derived from the Era Diocletian: the era counting "years of the World" (ʿāmata ʿālam)[6] or "since Adam", henceforth denoted by W, and the "era of the Incarnation" (śegāhu or similar terms) which begins 5500 years later than W:

(2) $$J\ 1 = W\ 5501.$$

The years of all these eras are Alexandrian years and begin with Thoth 1 = Maskaram 1. In particular we have for the era J of the Incarnation[7]

(3) $$J\ 1 \text{ Maskaram } 1 = \text{A. D. } 8 \text{ August } 29.$$

---

[4] E. g., Vat 61 (Mauro da Leonessa [1943] p. 323, 7r/326).

[5] "Years of Grace" for Diocletian: Jerus Arm 3483 193[b]—197[a]; Liber Axumae CSCO 58, pp. 81—83.

[6] "Years of Grace" also for W: Br. Mus. Cat. Wright p. 221 (BM 754 182[a]); CSCO 68, p. 4, 28; p. 131, 6. CSCO 138/139, p. 64/110 colophon.

[7] Also the years of the era J are sometimes called years of Grace (e. g., Berol 84, 1[a] or CSCO 61/66, p. 26/27).

On the basis of this convention one can obtain julian years from the years of Incarnation by adding 7 or 8 to the number of J. For example, J 277 corresponds to A. D. 284/5 (= D 1) and W 1 to the julian year —5492/1. Our Ethiopic documents, however, never refer to the julian era.

The subsequent Table I is based on the above given relations between the four eras W, J, D, and G. All transformations require nothing but additions or subtractions with respect to the year zero of the respective era. Column W, e. g., indicates that J = 0 corresponds to W = 5500, or D = 0 to W = 5776, etc. On the other hand line D implies that W 5776 = J 276 = D 0, etc.

Table I

Year 0 of	W	J	D	G_1	k
W	0	−5500	−5776	−5852	0
J	5500	0	−276	−352	180
D	5776	276	0	−76	456
G	5852	352	76	0	0
D_2	6308	808	532	456	456
G_2	6384	884	608	532	0
D_3	6840	1340	1064	988	456
G_3	6916	1416	1140	1064	0

This table shows clearly how this whole chronological system is based on the 19-year cycle with the Era Diocletian as a point of departure. Four 19-year cycles later begins the era G, 304 of these cycles earlier is the first year of the World. Thus the epochs of the eras W and G are separated from each other by an exact multiple (eleven) of the $28 \cdot 19 = 532$-year cycle. The number k in Table I indicates the position of the year 0 of each era with respect to this larger cycle.

Table II gives the julian day numbers for the epoch dates of the four eras. This suffices for the transformation of Ethiopic dates to any other contemporary chronological system, e. g. by using Schram's Tables of the Alexandrian calendar which agrees exactly with the Ethiopic calendar.

Table II

Era	Epoch			j. d.	jul. year N August 29	Weekday of Thoth 1
W	World	0 Thoth	0	9714 980	−5493	
		1	1	9715 346	−5492	Tuesday
J	Incarn.	0 Thoth	0	1723 855	A.D. 7	
		1	1	1724 221	8	Wednesday
D	Diocl.	0 Thoth	0	1824 664	283	
		1	1	1825 030	284	Friday
G_1	Grace,	0 Thoth	0	1852 423	359	
	cycle 1	1	1	1852 789	360	Tuesday
D_2	Mart.,	0 Thoth	0	2018 977	815	
	cycle 2	1	1	2019 343	816	Friday
G_2	Grace,	0 Thoth	0	2046 736	891	
	cycle 2	1	1	2047 102	892	Tuesday
D_3	Mart.,	0 Thoth	0	2213 290	1347	
	cycle 3	1	1	2213 656	1348	Friday
G_3	Grace,	0 Thoth	0	2241 049	1423	
	cycle 3	1	1	2241 415	1424	Tuesday

The history of these eras is quite clear, at least in the major outlines. The 19-year cycle governed the Mesopotamian lunar calendar since the fifth century B. C. and remained in use in Babylonian astronomy deep into the first century A. D.[8]. In the meantime Greek astronomers had adopted the far more convenient Egyptian years (of constant 365 days' length) which were replaced under Augustus by the "Alexandrian" years with a "julian" intercalation pattern. The astronomical eras, "Nabonassar" and "Philip" never became civil eras. This happened, however, to the regnal years of Diocletian, a development perhaps strengthened by the contemporary introduction of the taxation cycle known as the "indictio"[9]. At the same time a 19-year cycle regained

[8] Cf. Sachs [1976].

[9] Cf. below, p. 120.

calendaric importance through the Christian Easter computus[10], fashioned after the example of the Jewish Passover dates, which perpetuated a simple pattern for the determination of the lunar phases, based on the 19-year epact cycle.

It must have seemed natural to the Alexandrians to count the years of the Easter cycle in conformity with the contemporary civil era, i. e. the years of Diocletian. Thus it is not surprising to find both the Athanasian Easter dates (D 44 to 89) and the Ethiopic tables operating with an epact cycle that had its initial value $e = 30$ (Athanasius), or the equivalent $e = 0$ (Ethiopic), at D = 1.

An essential improvement was inaugurated by extending the cycle beyond the lunar phases in order to include not only the julian intercalations but also the weekday cycle. This meant the creation of a 532-year cycle[11], conveniently associated with the end of the fourth 19-year cycle, thus beginning with D 77. In all probability it was the monk Annianus (about A. D. 400)[12] who introduced this extended count of the cycle. Its initial year

(1) $$\text{D } 77 = \text{G } 1$$

is also the epoch year of the new "era" of Grace.

We also know the author of the construction of the two additional eras, the era since "Creation" (W) and the era of "Incarnation" (J): the monk Panodorus, a contemporary of Annianus[13]. In order to maintain the traditional speculations that Christ was born 5500 years after the Creation[14], and in order to associate Easter and Creation, Panodorus made the assumption that exactly 11 of the 532-year cycles had elapsed between Creation and G 1 = D 77. Hence G 1 became W $11 \cdot 532 + 1 =$ W 5853[15] and the Incarnation J 1 = W 5501. Neither the era J nor the era W is ever used in the tables on which the Easter computus is based.

Hence all characteristic elements of these four eras originated in Alexandria in the first century of the era Diocletian and remained the

---

10 Due to Anatolius, Bishop of Laodicea (around A. D. 270). Cf., e. g., Ginzel, Hdb. III, p. 232. This primitive procedure had nothing to do with the sophisticated methods of Babylonian astronomy.

11 Cf. s. v. Calendar, App. I (p. 56).

12 Cf. e. g., Ginzel, Hdb. III, p. 235.

13 Cf. Ginzel, Hdb. III, p. 179; Rühl, Chron. p. 191 f.

14 Cf. e. g., Schwartz, Ostert., p. 21.

15 It is meaningless to say "Christi Geburt fällt bei ihm [Pandorus] ins Jahr 5493" (Pauly-Wissowa, RE 18, 3 col. 633, 17). Pandorus was not concerned with an era invented more than a century after his time.

core of the Easter computus, as well as of Ethiopic chronology. In the West, however, the Alexandrian computus was accepted only after a long struggle by adopting, in the sixth century, the tables of Dionysius Exiguus[16]. His tables again are simply a copy of the Alexandrian tables as are the Ethiopic tables. Only in the counting of the era did he introduce a change by assuming that the epoch year of the previous era, D 1, corresponded to the year 284 after the birth of Christ, a norm which our present era still uses.

2. *Examples*

1. From a colophon[17] written on Hamlē 19 in the year

"World 7178, Grace 262, Incarnation 1678, Martyrs 1402".

The year Grace 262 obviously belongs to the third cycle, thus we may add $2 \cdot 532 = 1064$ years. This gives us the following equivalent dates, arranged in order

$$W\ 7178 \qquad J\ 1678 \qquad D_3\ 1402 \qquad G_3\ 1326$$

which shows the proper differences 5500, 276, and 76[18]. The day Hamlē (XI) 19 gives us as julian equivalence A. D. 1686 July 13 (July 23 greg.).

2. A traditional date for the "Conversion of Ethiopia" to Christianity is

J 245 = W 5745[19].

On the other hand we are told that

J 0 = year 180 in the 11th cycle[20].

Indeed, $10 \cdot 532 + 180 = 5500$ and $5500 + 245 = 5745$.

3. *Other Eras*

A. *Indictio*

A 15-year cycle, known as "indictio" (i) was introduced by Diocletian (presumably for the purpose of taxation), beginning in D 29 with $i = 1$. Thus we have

---

[16] Cf. s. v. Easter, App. II (p. 101).

[17] Dillmann, Bodl. Cat. No. 16, p. 29. For the continuation of this colophon cf. s. v. Calendar, p. 54.

[18] Cf. above Table I, p. 117.

[19] Berol 84 22[b] I, 27 f. The historicity of this date is very much in doubt and probably at least a century too early.

[20] Upps 3 63[b], 1 to 7.

(1a) $i\,0 = \mathrm{D}\ 28$ Thoth $1 =$ j. d. 1835257 (= A. D. 312 Aug. 29)

and consequently

(1b) $$\mathrm{D}\,0 \equiv i\,2 \mod 15.$$

Example: the year W 7000 has the indictio 11. Indeed, W 7000 = D 1224 = i 1226 and $1226 \equiv 11 \mod 15$.

On the basis of the relations between the four main Ethiopic eras (cf. above p. 117, Table I) one can easily establish rules for the direct determination of the indictio within each era. Thus for any year W and J one finds

(2a) $$i = \mathrm{W} - 5789 \equiv \mathrm{W} + 1 \mod 15$$

(2b) $$i = \mathrm{J} - 289 \equiv \mathrm{J} + 11 \mod 15.$$

For the eras D and G, however, one must take into consideration the 532-year cycles in which the dates in these eras are expressed. Then one has the following relations, valid within 532 years of the respective cycle:

Table III

Year 0 of	i	i (mod 15)	$i = 0$ for k
D_1	2	$D_1 + 2$	13
G_1	3	$G_1 + 3$	12
D_2	9	$D_2 + 9$	6
G_2	10	$G_2 + 10$	5
D_3	1	$D_3 + 1$	14
G_3	2	$G_3 + 2$	13

Example[21]: The year Diocletian 27 has the indictio $27 + 2 = 29 \equiv 14 \mod 15$. The same combination $k = 27$, $i = 14$ can only once more occur (in historical periods), namely for G_3 27[22]. Since no other parameter in the 532-year tables has a period of 15 years, only one of the two possible eras can be in agreement with the remaining calendaric dates.

The following Table IV makes it easy to establish the date and the era of any 19-year table which gives the indictio of its first year.

---

[21] From Jerus Arm 3483, 193^{b}—197^{a}.

[22] Because the difference $G_3\,0 - D_1\,0 = 1140$ is a multiple of 15.

Table IV

i		*k*					
G	D	1		2		3	
1	15	58	343	96	381	134	419
2	1	134	419	172	457	210	495
3	2	210	495	248		1	286
4	3	1	286	39	324	77	362
5	4	77	362	115	400	153	438
6	5	153	438	191	476	229	514
7	6	229	514	267		20	305
8	7	20	305	58	343	96	381
9	8	96	381	134	419	172	457
10	9	172	457	210	495	248	
11	10	248		1	286	39	324
12	11	39	324	77	362	115	400
13	12	115	400	153	438	191	476
14	13	191	476	229	514	267	
15	14	267		20	305	58	343
1	15	58	343	96	381	134	419

The three sections on the right part of the table belong to the cycles G_1, G_2, G_3 as well as to D_1, D_2, D_3. The numbers $k = 1, 20, 39$, etc. represent the first years of the 28 19-year cycles in each group. The left columns give the indictio i which corresponds to the numbers k listed in the same line. If k belongs to the era G, then i is to be taken from the first column, whereas D concerns the second column.

Examples: G_1 153 has $i = 6$ and the same i belongs also to G_1 438, G_2 191, etc., to G_3 514. Conversely: if a table begins with $k = 324$ and gives $i = 3$ the year and the era are uniquely determined as D_2 324 ($k = 324$ can otherwise only belong to $i = 12$ or 11 in G_1 or D_1, to $i = 4$ in G_2, and to $i = 11$ or 10 in G_3 or D_3).

In spite of the fact that the cycle of indictions has no astronomical meaning and that its economic significance must have been lost long before, Dionysius Exiguus listed the indictions of all years in his Easter tables[23], thus introducing an important chronological parameter into documents of the Latin Middle Ages[24]. In the Ethiopic tables indictions are relatively rare but nevertheless attested until the latest periods[25].

---

[23] Cf. s. v. Easter, App. II, p. 101.

[24] Cf. Krusch [1937], p. 59.

[25] E. g. BN Abb 123 184ᵃ—197ᵇ, a 532-year table for G_3. Beginning with $k = 305$ the values of i are repeated from $k = 286$ on and vitiate all remaining tables.

Tables for indictions are found, e. g., in Ginzel, Hdb. III, p. 393 ff., or in Grumel, Chron. p. 314[26]. Cf. also above s. v. Calendar, Tables p. 14 and s. v. Easter, App. I, p. 100.

B. Alexander

Besides the era of Diocletian the Seleucid Era (in its western version with years beginning with Tishre ∼ October −311) also was occasionally used in Ethiopic texts under the name of year of "Alexander the Greek" or "Alexander of Macedon". Apparently October fell near enough to the beginning of the Ethiopian years that one simply identified the years of "Alexander" with Ethiopian years.

Thus we find, e. g.[27], the relations

$$W\ 7074 = J\ 1574 = \text{Martyrs (D)}\ 129 = \text{Alex.}\ 1893.$$

Consequently[28]

$$\text{Alex.}\ 319 = J\ 0 \qquad \text{or} \qquad \text{Alex.}\ 0 = J-319 = W\ 5181$$

which indeed corresponds to julian −311/10 as epoch year.

C. Bizan

A provincial Ethiopic era takes its name from the famous monastery of Bizan (near Asmara). It is attested, e. g., in the text BM 754[29] in the form

$$(4a) \qquad B\ 6674 = W\ 7130.$$

If we add to the equivalent relation

$$(4b) \qquad B\ 0 = W\ 456$$

on both sides ten 532-year cycles, we obtain

$$(5) \qquad B\ 5320 = W\ 5776 = D\ 0.$$

This shows that the "Era Bizan" begins exactly ten great luni-solar

[26] Misprint in Table B: 87 instead of 78.

[27] Frankf 9 (Goldschmidt p. 23); Vindob 6 (Rhodokanakis [1906] p. 6). Similarly Berol 84 18ᵃ II, 22. CSCO 20 p. 100, 1 (21 p. 114) equates the next year with Alex. 1898 (instead of 1894). CSCO 20 p. 142, 2 (21 p. 162) reveals an error of −24 years for Alexander, of −4 years for D.

[28] Similarly Princ 5884 3ᵇ II: Alex. 595 = J 276. Also Ludolf, Comm. ad Hist. Aeth. (1691) p. 388, 16.

[29] Catal. Wright No. 331 p. 221ᵇ, BM 754 fol. 182ᵃ; also BM Add 16220 5ᵇ II, 2 (Dillmann. B. M. Cat. p. 10 No. 11) but correct 6757 to 6657. Another occurrence (Vat. Cat. p. 485, No. 119, fol. 90ᵇ) is unfortunately corrupt. Chaine, Chron. p. 115, n. 3, mistook Bizan for Byzantium, and so did Dillmann. The correct interpretation was given by Mauro da Leonessa [1943] p. 315, n. 1.

cycles before the Era Diocletian or, in other words, the cycles of the Era Bizan are identical with the cycles of the Era Diocletian[30].

D. Judith

A whole cluster of eras is mentioned in another text of the British Museum[31], always in relation to W 7113 (= J 1613 = A. D. 1620/21). The names are strange: the years of the World are "Era (ḥasāb) of the Apostles" as well as "Years of Grace". An "Era Judith" is simply equivalent to the cycle D_3. An "Era of the Martyrs" begins $38 = 2 \cdot 19$ years later than D, the "Era of the Justs" corresponds to Bizan + 400, the "Era of the Elders" to W − 400. An "Era of Rome" begins $300 \cdot 19$ years earlier than the above mentioned "Era of the Martyrs". All this gives the impression of ad hoc inventions.

E. Era 1255

A set of tables[32] for three (probably originally six) 19-year cycles begins with a year "1255". From the values of the parameters listed (tw, ε, and e which have a common smallest period of 532 years) one finds $k = 343$ as cycle year in G. Since $1255 \equiv 191 \bmod 532$, we can say that these years are counted from a point $343 - 191 = 152 = 2 \cdot 76$ years later than the epoch of the era G[33]. No such era (with the epoch J 505) seems to be attested elsewhere[34].

F. Hebrews

An "Era of the Hebrews", otherwise unknown, is connected with the Ethiopic era mundi[35] by

$$\text{year of the Hebrews } 6642 = \text{W } 7074.$$

Thus we have

$$\text{year } 0 = \text{J} - 5066$$

for the epoch, preceding the known Jewish era mundi[36] by 1300 years.

[30] This relationship is confirmed by another equivalence mentioned in the same text: B 290 = G 214. Hence the cycle numbers of B are 76 years higher than the cycle numbers of G, a relation which is also valid for D and G.

[31] BM Add 16220 5^b II, 1—8 (Catal. Dillmann, No. 11, p. 10/11).

[32] Ṭānā 34 $2^{a,b}$.

[33] The era G is used normally in the tables of the same text ($1^{a,b}$; 3^a to 16^b; 29^a to 92^b).

[34] The Easter tables of Cyrillos (Patriarch of Alexandria around 440) began with the year D 153, i. e. G 76 + 1. (Cf. Krusch [1937] p. 59; also Easter App. II, p. 101.)

[35] Goldschmidt, Frankf. No. 9, p. 23.

[36] Beginning −3760 Oct. 6/7. (Ginzel, Hdb. II, p. 79.)

G. Hijra

It is not surprising to find in Ethiopic texts references to the Islamic era. An unexpected and, at least to my knowledge, unique feature is a counting of the Islamic years (ʿāmata ʾagār) in two ways: either in the form of the actual Islamic calendar (based on "lunar" years of 354;22 days) or as julian years. This is evident in the colophon of the translator of the Chronicle of John, Bishop of Nikiu, where we find the following dates[37]: W 7594[38], Alexander 1947[39], J 1594, D 1318, and "980 Islamic years in solar years, 1010 years in lunar years". Indeed the ratio of 365.25 to 354.367 is very nearly 1.0307[40] and 980 times this factor is 1010.1. Consequently we have to distinguish two eras, both beginning with the Hijra; one we denote by A using julian years, the other AH, which follows the actual Islamic reckoning. For A we have the epoch

$$A\ 0 = W\ 6114 = J\ 614$$

most likely counted from Thoth 1. For the era AH the traditional epoch date is A. D. 622 July 15.

This ambiguity, combined with the fact that there exists no simple relation with constant differences for the transformation from Alexandrian to Islamic dates, perhaps explains the frequency of incorrect Hijra dates in our texts[41].

As examples for accurately computed Islamic dates, we may quote from Abū Shāker's Computus, ch. 15[42]:

D 972 Yakātit (VII) 3 Saturday = A. H. 654 Muḥarram (I) 1
(indeed, both dates correspond to A. D. 1256 Jan. 29 =
j. d. 2179840, Saturday)

and from the History of Girgis ibn ʿAmīd[43]

D 1011 Maskaram (I) 1 = A. H. 693 Šawāl (X) 7
(indeed, both dates correspond to A. D. 1294 Aug. 29 =
j. d. 2193932).

---

37 Zotenberg [1883] p. 342/587.

38 Error for 7094; cf. J 1594.

39 Should be 1913.

40 Sexagesimally 365;15/354;22 ≈ 1;1,50,33. The inverse ratio is 0;58,12,44 ≈ 0.972.

41 Vat Borg 1 3ᵇ, 1—15: Alex 1893 ≈ A. H. 909; Dillmann, Bodl. Cat. No. 16 p. 20: W 7178 = D 1402 ~ A. H. 1039; Vat 33 (Cat. p. 159, 5): J 275 + 320 = D 320 ≈ A. H. 1 (i. e. A. D. 602/3, instead of 622/3).

42 Dillmann, BM. Catal. p. 42.

43 Conti Rossini [1915] p. 194, No. 210. His "emendations" are caused by incorrect correspondences of his own making.

It may be noted that the Islamic day numbers in both cases follow the norm of the astronomers[44].

The "Guide for the Blind" (in a version written in 1896) gives also, besides the usual "solar years" of the era W, "lunar years" ('āmata warḫ), in particular for the life of Jesus[45]. Obviously these lunar years were obtained by multiplying the numbers of solar years with a factor corresponding to the ratio 365/354 (or similar) though the results are not very accurate[46]. W 5500, e. g., should correspond to about 5670 lunar years, whereas the text gives only 5667 and similarly for the remaining dates[47].

H. The Era W′ = W 38

The beginning of the era W with Tuesday ($j = 3$) agrees with Genesis in so far as the luminaries were created on the third day. Somebody must have felt, however, that a true "Era of the World" should begin with the first day of Creation, i.e. with a Sunday ($j = 1$). A glance at the tables shows that this is the case for $k = 39$, only two 19-year cycles later than the first year of W. Hence, by introducing a new era of the World (W′) which satisfies the relation

$$\text{(1)} \qquad \text{W}'\,0 = \text{W}\,38 \equiv \text{G}\,38 \quad \text{mod } 532$$

one counts also weekdays from the beginning of the world.

Cycle numbers k' of this new era, 38 lower than the cycle numbers k of W, lead, however, to differemt Easter dates. But if k is a proper date (e.g. $k = 214$ for $f =$ VII 29, the date of Resurrection in W 5534[48]) then the same date will be obtained for the four dates $k - 95$, k, $k + 95$, $k + 190$, i.e. for $k' = -57$, 38, 133, and 228. Among these four possibilities only the last one, $228 = 12 \cdot 19$, is a number divisible by 4, hence providing the proper position (ε) within the intercalation cycle. Thus, if one counts the cycle years in the Era W′ from a year H such that

$$\text{(2)} \qquad \text{H}\,0 \equiv \text{G}\,228 \quad \text{mod } 532$$

one will obtain the same easter dates as in the era G. For example for the date $f =$ VII 29 one has (mod 532)

---

[44] Cf. Neugebauer, Al-Khwār. p. 10, Fig. 1.

[45] Marha 'E 379 I, 23—380 I, 18.

[46] The text apparently uses the factor 1.03 instead of 1.031.

[47] The text gives also lunar dates (čaraqā, i. e. "luna") for the civil dates (but no month-names).

[48] Cf. p. 171.

H	G	ε
423	119	L.
518	214	Mr.
81	309	M.
176	404	J.

This peculiar chronological system is attested in several treatises, most extensively in BN 160 (24^a I, 19 to 24^b I, 3 and 25^b I, 2 to 27^a II, 21 for the chronology of Christ, and in general from 16^b to 27^a).

Evangelists

Each Ethiopic calendar year is associated with the name of one of the Evangelists, cyclically repeated in the canonical order:

(1) Matthew Mark Luke John.

For the corresponding computational rules, cf. s. v. Calendar p. 44 (ε).

We also have lists which arrange calendaric parameters according to the Evangelists of the corresponding years, covering $4 \cdot 19 = 76$ years. One such list concerns the epact[1], another text gives the distribution of the weekday of Maskaram 1, i. e. "Day of John" (j) and the equivalent "ṭentyon" (t)[2].

If it were not for the julian intercalation, all calendaric parameters would have a period of $7 \cdot 19 = 133$ years. It is therefore not surprising to find a division of the 532-year cycle into four groups of 133 years, each of which is associated with an Evangelist[3]. Since each of these groups contains seven 19-year cycles, the name "week of Matthew" (etc.) is given to them. In each "week" (sanbat) the individual 19-year tables are counted as "sections" ('anqaṣ) from 1 to 7. At the end of each group we read "this is the seat (manbara) of Matthew" (etc.).

References to these "weeks" are also found in marginal notes of 532-year tables at $k = 1$, 134, 267, and 400 (era G)[4]. In other 532-year

[1] BM Add 16217 17^a II, 10—17^b I, 6; Princ 5884 7^b II, 5—III, 1; Vindob 6 18^a II, 1—11.

[2] Vindob 6 21^a II, 1—13 (in part badly garbled).

[3] EMML 215 40^b I, 31—42^b II, 24; 43^a II, 20—44^b I, 27.

[4] EMML 215 83^a, 86^b, 90^a, 94^b; Vindob 6 4^a, 7^b, 11^a, 14^b.

tables, however, which begin with the year $k = 153$[5], we find the "beginnings" of the weeks of the Evangelists associated with $k = 153$, 286, 419 and 20, thus one 19-year cycle later than in the other tables. Furthermore, "ends" of these weeks are noted at $k = 283$, 416, 17, and 150, i. e. three years before the next "beginning"[6]. The reason for this anomaly lies probably in the fact that $c = 17$ in every 19-year table is associated with the same Evangelists as $c = 1$[7].

An entirely different meaning has the expression "years of the Evangelists", which refers to the 84 years[8] which are needed for the completion of 7000 years after 13 532-year cycles, i. e. after W 6916.

Another use of the term "years of the Evangelists" is also associated with the 7000 years, i. e. with the second coming of Christ: each fourth of this time, hence 1750 years each, is related to one of the Evangelists[9]. In general every tetrad can suggest similar connections. We find, e. g., the rivers of the Paradise and the Evangelists

Matthew	Euphrates
Mark	Giyon
Luke	Tigris
John	Fison

listed in the four corners of some cosmological diagrams[10].

[5] For this year as starting point, cf. s. v. Cycles p. 88.

[6] BM Add 16217 2^a—6^b I (with some omissions).

[7] EMML 215 40^b I, 31—42^b II, 24.

[8] Grébaut [1920] p. 219; BM Add 24995 27^a I, 13 f.; Princ 5884 12^a I, 5—7; Upps 3 57^b 6/7; 59^a 13/14; Vindob 6 6^a at $k = 84$. Cf. also s. v. Jubilees, p. 173.

[9] Grébaut [1920] p. 220; Upps 3 61^a 7—13.

[10] BM Add 16217 24^a; Jerus Littm. [1902] 3.

Evening Epoch

See Feasts, Lunar Dates (p. 139).

Ezra, Week of

See Jubilees (p. 172).

Feasts

See also Calendar (p. 27).

Table of Contents

1. General Outline

The moveable feasts[1] of the Ethiopic calendar fall into two sharply distinct groups:

A: Jewish festivals, in fixed relation to Passover (*p*)

B: Christian festivals, in fixed relation to Easter Sunday (*f*).

The determination of the dates of feasts belonging to either one of these two classes follows different sets of rules, as is inevitable since B involves the days of the week, but A does not.

---

[1] Feasts of fixed calendar date (e. g. "Michael" on the 12th of each month) are of no concern to the "Computus". A list of such feasts (from Ludolf) is given in Mauro da Leonessa, Diss. Cron. p. 135—151.

There exist other basic differences: dates can refer to the "civil calender", which begins with Maskaram 1 = Thoth 1[2], or to a schematic lunar calendar which begins with the Jewish New Years Day. Definite rules govern the transformation of one of these norms to the other.

Finally, dates may be given only by the day number within a certain month without naming the latter. This means a variability modulo 30 for dates in the civil calendar. The same procedure can also be followed for the intervals between dates so that, e. g., an interval of 190 days may be reduced to a difference of 10 for the day numbers which, however, belong to two months half a year apart.

We now return to the fundamental feature common to both groups A and B: the dates of all moveable feasts are known as soon as the date for one of them is found. The "Computus" makes use of this fact by deriving the dates of all feasts from one date early in the year as follows:

A. The basic parameter for the Jewish feasts is the maṭqeʿe, i. e. the civil date (m) of the Jewish New Years Day. This day always belongs to the civil months I or II, the date being determined by the epact (e) and $e + m = 30$. From m one obtains the date of Passover (p) by

$$(1) \qquad p = m + 190 \qquad \text{or} \qquad p \equiv m + 10 \mod 30.$$

B. For the Christian feasts the days of the week are involved, depending on the ṭentyon (t) which represents the weekday of the civil New Years Day, Maskaram 1. Then, on the basis of certain rules[3], the date of a feast called Nineveh (n) is determined from t and from the cycle number c of the year within the 19-year cycle. Finally, the dates of all subsequent feasts are at fixed distances from n. In particular Easter Sunday is given by

$$f = n + 69 \qquad \text{or} \qquad f \equiv n + 9 \mod 30.$$

Remark. For the sake of simplicity I always call "feasts" what are sometimes distinguished as "fasts and feasts" in the texts.

2. *Spacing of Feasts*

In enumerating the distances of feasts from the dates of m or n, respectively, our texts vary freely between "absolute" intervals, which represent the actual distance in days, and "reduced" intervals, which, by reductions modulo 30, provide the day numbers alone, taking the

---

[2] Excepting the names of the months the Ethiopic civil calendar is identical with the Alexandrian calendar, introduced by Augustus.

[3] Cf. s. v. Calendar, Parameters, p. 48.

months for granted. It is important to remember that an addition to a reduced date does not necessarily imply transition to a later date. Consequently the numerical order of reduced dates need not express actual chronological order; for example, $mo = n + 11$ falls at a later date than $bf = n + 14$. Similarly subtractive terms always can be changed to additive reduced dates. For example, $-10 \equiv +20 \mod 30$.

The feasts ennumerated in the following lists represent what is usually discussed in the Ethiopic calendar treatises. For details cf. the notes in Section 6 (p. 142 ff.).

	A. Jewish Feasts	absolute	reduced
m	Maṭqeʿe, i.e. New Year	m	m
yk	Yom Kippur (ṣoma ʾayhud or badr)	m + 9	m + 9
tb	Tabernacle (maṣalat)	m + 44	m + 14
p	Passover (feśeḥ)	m + 190	m + 10
pu	Purim, i.e. Fast of Salvation (ṣoma deḫnat)	m + 321	m + 21

The dates of all Jewish festivals decrease (modulo 30) by 11 days from year to year[4]. This allows us to establish simple checking rules in relation to the epact e:

(1) $$e + m = 30$$

(2) $$e + p = \begin{cases} 10 & \text{for } e < 10 \\ 40 & \text{for } e > 10 \end{cases}$$

(3) $$e + tb = \begin{cases} 14 & \text{for } e < 14 \\ 44 & \text{for } e \geqq 14 \end{cases}$$

We also find frequently short rules for determining from a given date of a Jewish festival (say π_0) its dates in subsequent or preceding years ($\pi_1, \pi_2, \ldots$ or $\ldots \pi_{-2}, \pi_{-1}$ respectively). Since one always operates with numbers modulo 30, one can make all differences positive. Thus one has[5]

---

[4] Cf. "Table XIX" s. v. Calendar p. 59.

[5] I ignore here the one exceptional difference caused by the saltus lunae. Our texts do the same.

$$(4)\quad \begin{matrix} \pi_1 = \pi_0 + 19 & \pi_2 = \pi_0 + 8 \\ \pi_{-1} = \pi_0 + 11 & \pi_{-2} = \pi_0 + 22 \end{matrix} \quad \text{etc.}$$

Similar rules combine a feast of one year with different feasts in a subsequent or earlier year[6]. It follows, e. g., from

$$p = m + 10 \qquad \text{and} \qquad tb = m + 14$$

that $tb = p + 4$. Therefore with (4) $tb_1 = tb_0 + 19$ and thus

$$(5) \qquad tb_1 = p_0 + 23.$$

In the same way the dates of all Jewish feasts in future or past years can be determined from one Passover date.

Here, as often in the following, I use the equal sign instead of writing congruences modulo 30. This shortened terminology corresponds also to the formulations found in the texts.

B. Christian Feasts		absolute	reduced	weekday
n	Nineveh	n	n	Monday
bf	Beginning of Fast (ba'ata ṣom)	$n+14$	$n+14$	Monday
mo	Mount Olive (dabra zayet)	$n+41$	$n+11$	Sunday
ho	Hosanna (hosā'ena)	$n+62$	$n+2$	Sunday
f	Easter (fāsikā)	$n+69$	$n+9$	Sunday
ra	Synod (rakeb)	$n+93$	$n+3$	Wednesday
asc	Ascension ('ergat or ba'ala 40)	$n+108$	$n+18$	Thursday
pt	Pentecost (ba'ala 50)	$n+118$	$n+28$	Sunday
md	Prayer of Salvation (meḥlā deh̲nat)	$n+121$	$n+1$	Wednesday
sh	Fast of Apostles (ṣoma ḥawāryat)	$n+126$	$n+6$	Monday

Since Easter (*f*) was made (after much bitter struggling) to be always a Sunday, the weekdays of all other moveable Christian feasts must fall always on the same weekday because their distances from *f* are fixed.

Two texts in a group of five (generally parallel texts) make $sh = n + 119$, thus placing the feast on the Monday after Pentecost and thus before *md* on Wednesday[7].

---

[6] Berol 994 67ᵃ I, 19—67ᵇ I, 10.

[7] BM Add 24995 24ᵃ I, 19 f., II, 6 f., 24ᵇ I, 15—19, II, 18; Vat 119 87ᵃ II, 15 f., 87ᵇ I, 11 f., II, 5—8 (in contrast to BM 816; EMML 215; Ṭānā 34; Vindob 6 which follow the tradition $sh = n + 126$). Cf. also below p. 133 and p. 151.

3. Calendaric Boundaries of Feasts

Since all feasts in A as well as B maintain fixed distances from m and n, respectively, all feasts in both groups must range over the same interval that is permitted for m or n. This is in fact the case in the corresponding rules found in our texts. Accordingly Jewish feasts have 29 days (end-points included) at their disposal, i. e. one lunar month. The Christian Feasts, however, vary over 35 days (end-points included), i. e., five weeks, in consequence of the fixed association with the days of the week.

The following list gives the permitted intervals:

A:
I 15 $\leqq$ maṭqeʿe (m) $\leqq$ II 13
I 24 $\leqq$ badr (yk) $\leqq$ II 22
II 29 $\leqq$ Tabernacle (tb) $\leqq$ III 27
VII 25 $\leqq$ Passover (p) $\leqq$ VIII 23
XII 6 $\leqq$ Fast of Salvation (pu) $\leqq$ epag. 4

B:
V 17 $\leqq$ Nineveh (n) $\leqq$ VI 21
VI 1 $\leqq$ Beginning of Fast (bf) $\leqq$ VII 5
VI 28 $\leqq$ Mt. Olive (mo) $\leqq$ VIII 2
VII 19 $\leqq$ Hosanna (ho) $\leqq$ VIII 23
VII 26 $\leqq$ Easter (f) $\leqq$ VIII 30
VIII 20 $\leqq$ Synod (ra) $\leqq$ IX 24
IX 5 $\leqq$ Ascension (asc) $\leqq$ X 9
IX 15 $\leqq$ Pentecost (pt) $\leqq$ X 19
IX 18 $\leqq$ Thanks for Salvation (md) $\leqq$ X 22
IX 23 $\leqq$ Fast of Apostles (sh) $\leqq$ X 27

If one assumes $sh = n + 119$, then one has IX 16 $\leqq sh \leqq$ X 20.

As will be shown in the next section, the limits in B are a consequence of the limits in A.

4. Connections between Jewish and Christian Feasts

At the core of the whole Ethiopic, i. e. Alexandrian, Easter Computus lies the norm: Easter-Sunday is the first Sunday after Passover. Or in other words: Easter Sunday can come as close as one day to Passover and falls at most one week after Passover.

That this is the governing rule of the Computus can be demonstrated not only from a scrutiny of the actual 532-year tables but also from rules in our treatises that postulate various relations between Jewish and Christian feast days. We shall present both aspects in the following discussion.

Table I

c	p	f						
1	10	11	12	13	14	15	16	17
2	29	30	1	2	3	4	5	6
3	18	19	20	21	22	23	24	25
4	7	8	9	10	11	12	13	14
5	26	27	28	29	30	1	2	3
6	15	16	17	18	19	20	21	22
7	4	5	6	7	8	9	10	11
8	23	24	25	26	27	28	29	30
9	12	13	14	15	16	17	18	19
10	1	2	3	4	5	6	7	8
11	20	21	22	23	24	25	26	27
12	9	10	11	12	13	14	15	16
13	28	29	30	1	2	3	4	5
14	17	18	19	20	21	22	23	24
15	6	7	8	9	10	11	12	13
16	25	26	27	28	29	30	1	2
17	14	15	16	17	18	19	20	21
18	3	4	5	6	7	8	9	10
19	22	23	24	25	26	27	28	29

We will begin with an inspection of the 532-year tables[8]: in Table I the dates of Easter (f) are given as a function of the 19 possible dates (p) of Passover[9]. This table makes it evident that f is restricted to the week from $p+1$ to $p+7$, and therefore contains only one Sunday, depending on the weekday of p. This pattern contains $7 \cdot 19 = 133$ dates. In 532

[8] Cf. above, s. v. Calendar, p. 59—63.

[9] Underlined dates belong to month VII, all other dates to VIII. The earliest dates are associated with $c = 16$, the latest with $c = 8$.

years these dates are repeated four times at intervals of 95 years (mod 532)[10]. It is also evident from Table I that the limits for Passover

$$\text{VII } 25 \leqq p \leqq \text{VIII } 23$$

imply for f the limits $25+1 = 26$ and $23+7 = 30$, respectively, thus

$$\text{VII } 26 \leqq f \leqq \text{VIII } 30$$

for Easter and similarly for all moveable Christian feasts[11].

Secondly, we have rules that connect the dates of Christian and Jewish festivals, depending on the weekday on which the latter fall. Since all Jewish feasts are known when the New Years date m is known, and all Christian feasts when Nineveh (n) is known, the following rule which relates n to m is of special, practical importance[12]:

(1)
if m falls on a Saturday, then $n \equiv m+8 \mod 30$.
Sunday $m+7$
etc.
Friday $m+2$

It is easy to show that these relations are a direct consequence of the rule that Easter is the first Sunday after Passover. Indeed:

(2a)
if m falls on a Saturday, then p is a Sunday
Sunday Monday

etc., because $p = m+190 \equiv m+1 \mod 7$. Now the rule for Easter Sunday implies that

(2b)
for p on a Sunday f is the Sunday $= p+7 = m+197$
Monday Sunday $= p+6 = m+196$

etc. Therefore, from (2a) and (2b) and p. 132B:

(2c)
if m is a Saturday $n = f-69 = m+128 \equiv m+8$ mod 30
Sunday $+127 \equiv m+7$

etc. Q. e. d.

---

10 Cf. s. v. Ṭentyon, Appendix, p. 223.

11 Cf. above, p. 133.

12 Preserved in many treatises, e. g., Berol 84 9^b I, 29—II, 4; Bodl 21 1^a I, 18—II, 3; Princ 5884 6^a I, 3—5; Upps 3, 72^b, 8—14; Vat 119 86^b II, 19—87^a I, 4.

13 Princ 5884 11^a I, 4—12. Similar rules in BM Add 24995 23^b II, 13—21.

Some treatises give also absolute intervals between Jewish and Christian feasts, depending on the weekday of m. For example[13]:

$$(3)\quad \begin{array}{lll} \text{if } m \text{ falls on a Saturday then } & bf = & m+142 \\ \text{Sunday} & & m+141 \\ \text{etc.} & & \\ \text{Friday} & & m+136 \end{array}$$

which leads indeed to (1) since $bf = n + 14$.

Conversely: it follows from (1) that for m = Saturday

$$f = n + 69 \equiv m + 8 + 69 \equiv m + 17 \mod 30$$

and therefore, since always $p \equiv m + 10 \mod 30$

$$(4)\quad \begin{array}{lll} f = p+7 & \text{if} & m \text{ falls on a Saturday} \\ p+6 & \text{etc.} & \text{Sunday} \\ p+1 & & \text{Friday} \end{array}$$

which tells us that Passover varies between $f - 7 = ho$, i. e. Palm Sunday, hosāʿenā, (if the Jewish New Years Day is a Saturday) and the Saturday before Easter (if m = Friday), the Easter Vigil, māḫatwa fāsikā[14].

Since the lunar date p of Passover is, by definition, always 14, it follows from (4) that the lunar dates of Easter Sunday are limited by

$$(5)\quad 15 \leqq \bar{f} \leqq 21.$$

It is of interest to note that VII 26 as the lowest civil date of f and 21 as the upper lunar date (i. e. "luna" in medieval terminology) are mentioned in as early a document as the "Didascalia" (third century A. D.). These limits imply the same range for f as shown in Table I as soon as the dates of p are determined by the epact scheme of the 19-year cycle. Since there is no trace of any other Jewish calendar pattern, it seems clear that the Alexandrian Easter computus from its early days located Easter in the week following Passover.

The political background of this rule, of course, lies in the desire to avoid any coincidence of Easter Sunday with Passover. If furthermore one defined VII 25, i. e. the earliest day permitted for Passover, as the date for the vernal equinox[15] one would have at the same time the guarantee that Easter would always follow the equinox. Thus the entire Easter computus reveals itself as an extremely simple arithmetical pattern, both for Passover and for Easter Sunday, without any consid-

---

[14] This formulation is found in Vat 88 50[a] I, 8—II, 1.

[15] Cf. below s. v. Seasons, p. 205.

eration of astronomical concepts whatsoever. But it must be admitted that this arithmetical simplicity conveys a certain elegance to the whole structure upon which the 532-year table is based.

5. *Lunar Dates*

The 532-year tables contain for each of four Jewish festivals—New Years (m), Yom Kippur (yk), Tabernacle (tb), and Passover (p)—two columns, one for the civil dates, the second for the dates in a schematic lunar calendar of alternating full and hollow months. The civil dates vary in a 19-year cycle, controlled by the epact e, whereas the lunar dates are always the same within their lunar month. In particular the lunar date of Passover is always 14[16], indicating full moon, i. e. "luna 14" in medieval terminology.

Since we know that the 532-year tables represent the parameters of the Alexandrian Easter Computus at the end of the third century[17], it is also obvious that the lunar dates must refer to the contemporary Jewish Passover computus. Thus our tables confirm a conjecture of Ed. Schwartz that Jewish communities in the Diaspora regulated their lunar calendar within the framework of the contemporary local civil calendar[18]. This is indeed evident in the present case. The New Years Day m of lunar date 1 is given a civil date which oscillates between the dates I 15 and II 13, varying according to a simple arithmetical scheme so that the "epact" $e = 30 - m$ increases each year by 11 days (mod 30), except for one increase of 12 at the end (the "saltus lunae") in order to close the cycle every 19th year.

Such a pattern leads automatically to the addition of 7 "intercalary" months in 19 lunar years since $19 \cdot 354 + 7 \cdot 30 = 6936$ days as compared with $19 \cdot 365 = 6935$ days—the one-day difference being absorbed by the saltus lunae. Fig. 3. illustrates this situation[19]. The boundaries for m require the addition of 30 days each time a date would be too early. This happens seven times in 19 years:

---

16 Grébaut's statement in Vat. Cat. p. 484 "quod est XI per totum laterculum" is based on a trivial misreading 11 for 14, the canonical date of Passover.

17 Cf. s. v. Easter, App. I, p. 99.

18 Schwartz, Ostert. p. 126.

19 The upper edge in Fig. 3 represents the beginning of the civil years, numbered to conform with the cycle years of the eras W, D, and G. Time increases downwards. The points m and p recede 11 days from year to year, 12 days at the end. The points p are always 190 days distant from the points m in the same year.

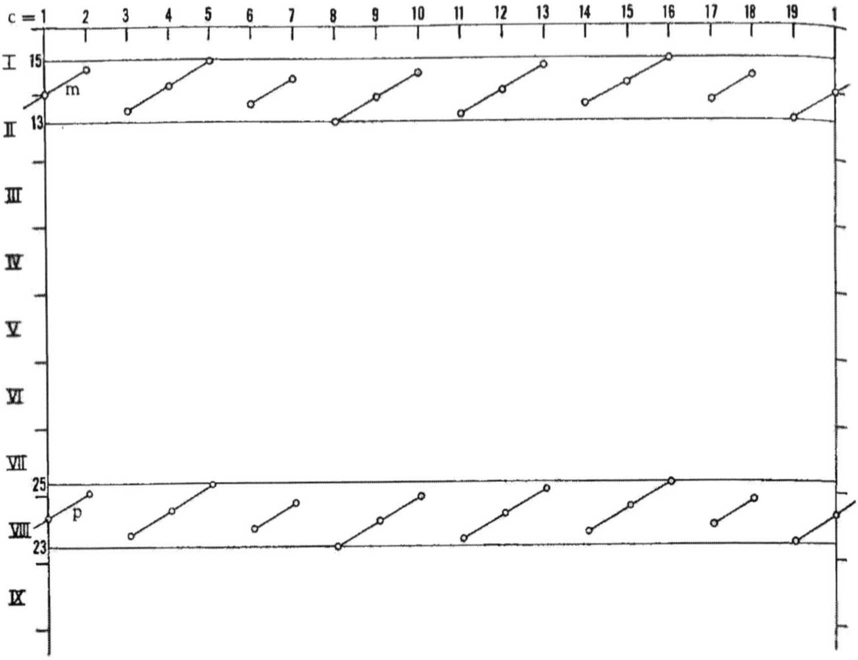

Fig. 3

$c = 2$	5	7	10	13	16	18	$21 \equiv 2$	mod 19
Δ:	3	2	3	3	3	2	3	years

This procedure singles out seven years as intercalary and determines their spacing Δ. Exactly the same pattern is well known from the Babylonian calendar which makes the following years of the Seleucid era intercalary[20]:

S.E. 4	7	9	12	15	18	$20 \equiv 1$	$23 \equiv 4$	mod 19
Δ:	3	2	3	3	3	2	3	years

The fact that the Alexandrian calendar counts six epagomenal days every fourth year is simply ignored for the lunar dates. Consequently, as we know from the 532-year tables, the dates of m (and thus of all Jewish feasts) always follow the same pattern in each 19-year cycle, regardless of whether the civil year is an ordinary or a leap year. Hence

[20] Cf. e. g., ACT p. 33. These years begin in Nisan, not Tišre. If one extends the 532-year cycle back to the (Eastern) Seleucid Era one finds that its first year ($c = 1$) would be $c = 5$ in the Alexandrian-Ethiopic count.

we know for all years the dates of the Jewish festivals in the civil calendar. Such dates are characterized in our treatises by the term śarqa maʿālt, literally "beginning of day", in contrast to the lunar dates, referred to as śarqa lēlit, "beginning of night". These terms represent in fact what we would call morning- and evening-epoch[21].

A drastic example of this terminology is the following passage[22]: "on the 28th śarqa lēlit the moon becomes invisible". This meant that on the 28th "of a lunar month" the moon becomes invisible, a phase that occurs in the morning, not at the "beginning of night" as one could (and did) naively translate.

It is important, however, to emphasize that this terminology does not imply a practical recognition of the different epochs of the day. All dates, civil or lunar, refer to the same arithmetical system of consecutive integers which nowhere introduce fractions of a day. This attitude is evident in such terminology as "luna 14" for Passover, followed by "luna 15" or "luna 16" for Easter.

We now wish to determine the lunar date $\bar{d}$ of a Jewish feast of given distance δ days from the New Years Day $m = 1$. If all months were 30 days in length, this date would be simply $1 + d$ in month $\mu + 1$ where $d = \delta - 30\mu$. We must remember, however, that we are dealing with schematic lunar months of alternating 30 and 29 days. Each completed pair of such alternating months, i. e. each complete interval of 59 days, raises the day number in the next month by 1. Hence we have

(1) $$\bar{d} = 1 + d + [\delta/59]$$

where $[\delta/59]$ is the number of pairs of full and hollow months contained in δ[23]. In practice $[\delta/59]$ will be the same number as $[\mu/2]$ where μ is the number of completed months[24]. Hence we have simply

(2) $$\bar{d} = 1 + d + [\mu/2] \qquad \text{with } \delta \equiv d \mod 30$$

for the lunar date in month $\mu + 1$.

---

21 Hence we see that the Ethiopic calendar operates with days in morning epoch, thus following the Alexandrian norm. Also the Coptic epoch for the days is sunrise (Sobhy [1942] p. 181).

22 BM Add 16220 5^b II, 11/12.

23 We use the notation $[n]$ for the greatest integer contained in n, e. g. $[3/2] = 1$, $[2/3] = 0$, $[63/59] = 1$.

24 Cf., e. g. Fig. 3, p. 138 for p where $\delta = 190$, thus $\mu = [190/30] = 6$ and $[190/59] = 3 = [\mu/2]$. In BN 160, 44^b II, 6—45^a II, 16 the pairs of consecutive hollow and full lunar months are counted as "manfaq" (lit. "halves") from 1 to 6.

We now apply this procedure to the Jewish festivals with their given distances from the maṭqeʿe[25]:

$$
(3)\qquad
\begin{array}{lrl}
\text{badr } (yk) & \delta = 9 & \\
\text{Tabernacle } (tb) & 44 \equiv 14 & \bmod 30 \\
\text{Passover } (p) & 190 \equiv 10 & \\
\text{Fast of Salvation } (pu)^{26} & 321 \equiv 21. &
\end{array}
$$

In the first two cases we have $\mu = 0$ or 1, hence $[\mu/2] = 0$ and the only amount to be added to 1 is the (reduced) number δ. Hence the lunar dates for yk and tb are simply

$$(4\text{a})\qquad \overline{yk} = 1 + 9 = 10 \qquad \overline{tb} = 1 + 14 = 15.$$

For the last two feasts we have $\mu = 6$ and $\mu = 10$ respectively; therefore $[\mu/2] = 3$ and 5, and thus for the lunar dates

$$(4\text{b})\qquad \bar{p} = 1 + 10 + 3 = 14 \qquad \overline{pu} = 1 + 21 + 5 = 27,$$

again in agreement with the conventional lunar dates in the tables[27].

It is inherent in the concept of "lunar dates" that these numbers can be interpreted as the "age" of the moon, in particular 14 as the "full moon", in good Babylonian tradition. Thus the date of Passover became the "luna 14", the Easter full moon, to medieval computists. If f is the date of Easter Sunday, then $\bar{f} = 14 + (f - p)$ is the "age" of the moon at Easter. Also, this luna is usually listed in medieval Easter tables[28] as a criterion for the distance between Easter and Passover.

A statement in the Ethiopic ḥasāb—"Easter in evening epoch does not go below 15 and does not exceed 21"[29]—confirms our previously obtained result[30] that Easter varies between $p+1$ and $p+7$. This is also characteristic for the Alexandrian Easter canon which permits a "luna 15" for Easter, a lower limit abhorred by the Latin church which insisted on a minimum of a "luna 16". This is one of the major sources of the Easter controversies, obviously totally unrelated to any underlying cycle or any other astronomical concept. In fact, astronomical accuracy or inaccuracy is of much less significance for the Easter computus than one would conclude from the rhetoric of modern church historians.

---

25 Cf. above p. 131.

26 Cf. below p. 144.

27 Also in treatises; cf., e. g., Wright, B. M. Cat. No. 312, p. 203 (BM 734 282 a I, 1—12).

28 Cf. s. v. Easter, App. II, p. 101.

29 Princ 5884 5 b I, heading; similarly 7 a III, 16 to 18.

30 Cf. above p. 134 Table I.

As we have seen, the lunar dates implicitly assume seven intercalary lunar months in each 19-year cycle. Excepting an occasional mention of a "14th of Nisan" (for Passover) the lunar months are rarely given names in the Ethiopic treatises[31]. Lunar dates mentioned in such texts either omit month names or retain the Ethiopic names for the civil month into which the day in question falls. For example, the lunar date "Nahasē (XII) 27" means a lunar month "XII" day 27, roughly coinciding with the civil month XII.

We also have lists of lunar months in their relation to the civil calendar (extending over 235 months, i.e. over one complete 19-year cycle)[31a] where the same names are used for lunar and for civil months. Thus we read[31b], e.g., "Ḫedār begins on the 19th of Ṭeqemt and ends on the 17th of Ḫedār", meaning that the lunar month III begins on the civil day II 19 and ends on III 17 (thus being a hollow month).

The general problem of finding the lunar date $\bar{d}$ of any civil date d is also occasionally treated in our texts[32]. For its solution one has to determine the distance of the given date from the date of m. We may assume as given for the date in question the epact e of the year. The value of e provides us, since $e = 30 - m$, with the number of days still left in the month to which m belongs. If the day in question falls into the civil month $\mu + 1$, the distance δ of its day d from m is either

$$\delta = e + \mu - 2 + d$$

or

$$\delta = e + \mu - 1 + d$$

depending on m being a day of month II or of month I respectively. Thus the lunar date will be

$$\bar{d} = 1 + [\mu'/2] + e + d \tag{5}$$

where $[\mu'/2]$ is either $[\mu/2] - 1$ or $[(\mu - 1)/2]$[33].

Occasionally lunar dates are sloppily determined. For example, we read in a colophon: "month Magābit (VII) on the 12th in evening (-epoch), on the 6th in morning (-epoch), in the year 1879 of the In-

---

[31] Jewish month names are, of course, well known as shown by many lists of foreign month names: Hebrew, Syriac, Coptic, etc.

[31a] BN 160 44^b I, 14—52^b II, 22.

[31b] BN 160 45^b I, 3—5.

[32] Princ 5884 5^b II, 7—12. Vat 61 6^a (Mauro da Leonessa [1943] p. 323/326). Both texts contain some slight errors of formulation.

[33] Strictly speaking one should use $[(\mu' \cdot 30 + e + d)/59]$ but only if $e + d$ combine to a full month would this influence the result.

carnation"[34]. Now J 1879 = G 463, hence $c = 7$, $e = 6$, $m =$ I 24. Hence the lunar date that corresponds to VII 6 should be

$$\bar{d} = 1 + 6 + [5/2] + 6 = 15$$

and not 12. Obviously the scribe simply made $\bar{d} = e + d = 12$, forgetting to correct for the intermediate 5 months and the initial 1, the lunar date of the initial day m.

6. *Notes on Specific Feasts*

A. *Jewish Feasts*

1. *Maṭqeʿe* (m)

Literally "trumpet", the Jewish New Years Day, therefore $\bar{m} = 1$. Cf. below s. v. Maṭqeʿe and Epact (p. 187).

2. *Yom Kippur* (yk)

Day of Atonement, lunar date $\overline{yk} = 10$. Called either ṣoma 'ayhud, "Fast of the Jews", or badr, "Account of life"[35]. The identity of these two feasts follows from many passages in our treatises, e. g.[36] "maṭqeʿe śarqa lēlit (i. e. $\bar{m}$) always falls on (day) 1, badr śarqa lēlit (i. e. $\overline{yk}$) always on (day) 10".

3. *Tabernacle* (tb)

Lunar date $\overline{tb} = 15$.

4. *Passover* (p) *and Feast of Harvest* (bs)

Ethiopic feśḥ, from Hebrew פסח, Greek πάσχα. The correct derivation of the latter term was given by Origines, correcting Irenaeus and Tertullian, who assumed a connection with πάσχειν, to suffer[37].

The lunar date of Passover is $\bar{p} = 14$. For the civil dates p and the weekdays pw, cf. s. v. Calendar p. 51 f. and Table "XIX" p. 59.

The precursor of the Christian "Pentecost", the Jewish "Feast of Harvest" (baʿāla sawit) is occasionally mentioned with its parameters[38]

$$bs = m + 240 = p + 50 \qquad \overline{bs} = 5.$$

---

[34] Toulouse No. 4, Chaine [1936] p. 15/16: ba 12 lēlit ba 6 maʿālt, omitting śarqa. This caused Chaine to translate "achevée . . . la douzième nuit, le sixième jour".

[35] Cf. Dillmann, Lex. 541 s. v. badr.

[36] Upps 3 71 ᵇ, 13—15.

[37] Greek text of Origines and French translation: Pierre Nautin, Homélies pascales II, p. 34 f. (Sources Chrétiennes 36, 1953).

[38] Marḥa 'E 375 I, 14 and I, 19; I, 2.

Since $pt = f + 50$ a coincidence of the two related feasts is as safely excluded as the coincidence of f and p.

5. Inner of Earth (*ce*)

This feast presents us with similar problems as do the two following, "Purim" (*pu*) and "Judith" (*ju*). In the ordinary 532-year tables these three feasts are not listed but they are included in enlarged versions[39]. From their position in these tables one would conclude that they belong to the latter part of the civil year, a conclusion which is confirmed by the arithmetical rules found in our treatises. Several features, however, point to a relation with a "year" that began in Nisan.

In some treatises the names of all three feasts are amplified. Thus Purim is the "Salvation from Haman by Esther" and Judith is the "Judith of the time of Holophernes". Similarly the Inner of Earth (leba meder)[40] is referred to as "the time of the Erythrean (Sibyl)"[41], probably because of her dwelling in a cave near Erythrai[42]. This is not the only connection with the Sibyls in our material[43]. The "Computus of the Sibyl" divides the 7000 years of the World into nine "suns" or nine "ages" (zaman), a terminology known also in connection with the Tiburtine Sibyl[44].

Turning to the arithmetical relations which define *ce*, it should be noted that all congruences are understood as modulo 30, if not explicitly stated otherwise. For the sake of easier comparison, relations to the maṭqeʿe (*m*) are indicated even if not directly mentioned in the texts.

For *ce* we have three groups of relations:

A. (1) $ce = yk + 5 = m + 14$[45]

(2) $\overline{ce} = 15$[46] or I 15[47]

[39] Cf. s. v. Calendar, Tables, p. 34.

[40] The same terminology is used in Matth 12, 40: ἐν τῇ καρδίᾳ τῆς γῆς.

[41] BM 815 24ᵇ II, 6 (and parallels: BM Add 24995, EMML 215, Vindob 6): 'ēretrā.

[42] On the Ionian coast, opposite Chios. For the ancient myths and their Jewish and Christian versions, cf. the article by Rzach in Pauly-Wissowa, RE II A 2, col. 2073—2183.

[43] Ḥasāba sābēlā (Grébaut [1920] p. 216); parallels and variants: BM Add 16217 17ᵃ upp. m., BM Add 24995 31ᵇ I, 11; Upps 3 60ᵇ, 10.

[44] Cf., e. g., Hennecke-Schneemelcher, NT Apocr II p. 708/9.

[45] Berol 84 10ᵃ II, 8 f.; EMML 215 62ᵇ II, 17—19.

[46] Berol 994 38ᵃ—65ᵃ; Princ 5884 6ᵃ III, 17—19; Vat 15 279ᵃ II, 12—16.

[47] Berol 994 66ᵃ I, 6—8; Upps 3 72ᵃ 6 f.

B. (3) $ce = p + 23 = m + 3$[48]
(4) VIII 18 $\leq ce \leq$ IX 16[49]
(5) $\overline{ce} = 7$[49].

The relation (4) is a consquence of the condition for m

$$\text{I } 15 \leq m \leq \text{II } 13$$

in combination with (3). Similarly (5) follows from (3) and (4) because one obtains for the lunar date[50]

$$\overline{ce} = 1 + 3 + [7/2] = 7.$$

The "Guide for the Blind" (version of 1896) has instead of B (3) and (5) the following relations: $ce = p + 22 = m + 2$ and $\overline{ce} = 6$[51].

Finally,

C. (6) $ce = p + 24 = m + 4$[49]
(7) $\overline{ce} = 8$[52].

Comments.

Since m increases from year to year by 19 group A (1) and B (3) could concern consecutive years. The same relationship will also be found between the groups A and B for *ju* (below p. 145). Also, the relation between (1) and (2) in A is the same for *ce* and *ju*. Lunar dates, however, should remain the same for all years. I cannot explain the decrease from 15 in (2) to 7 in (5) or 8 in (7). A similar decrease (from 10 to 4) is found for *ju*.

According to (4) *ce* should belong to the months VIII and IX, which would exclude $\overline{ce}$ = I 15 in (2). Perhaps "I" means Nisan here.

The dates in C are simply 1 day higher than in B (3) and (5), for reasons unknown.

6. *Fast of Salvation* (*pu*)

This feast, called ṣoma deḫnat, "Fast of Salvation", or ṣoma 'astēr, "Fast of Esther"[53], commemorates the "Salvation from Haman through Esther"[54], i. e., Purim. We are given the following relations:

---

48 Berol 994 38ᵃ—65ᵃ; BM 816 1ᵃ—5ᵃ.
49 EMML 215 6ᵇ II, 21—28.
50 Cf. above p. 139.
51 Marḫa 'E 374 II, 6—8; 375 I, 1, I, 13, I, 18.
52 BM 816 1ᵃ—5ᵃ.
53 Guidi [1896] p. 373 II, 26 etc.
54 BM Add 24995 20ᵇ I, 21; 25ᵃ I, 20; EMML 215 2ᵃ II, 3—4; Vindob 6 23ᵃ II, 9—10. Cf. also Mauro da Leonessa [1943] p. 306.

(1) $pu = m + 321 \equiv p + 11 \equiv m + 21$[55]
(2) $\text{XII } 6 \leq pu \leq \text{ep } 4$[56]
(3) $\overline{pu} = \text{XII } 27$[57].

It is clear from these data that the feast falls near the end of the civil year, i. e. August, contrary to the Biblical date of Purim[58], Adar 14, i. e., one month before Passover. This discrepancy speaks strongly for a shift from a reckoning from Nisan to an autumnal epoch of the year, a shift which is well attested e. g. in the counting of the seasons[59].

The boundaries (2) are consistent with (1) and the boundaries for *m* or *p*. Also the lunar date (3) is correctly determined. The relations between (2) and (3) are the same as for the corresponding data in group B of *ce* and *ju*.

7. *Judith* (*ju*)

This feast is called "Fast of Judith, by Judith in the time of Holophernes"[60], or simply "Fast of Judith"[61] or "Judith". As in the case of *ce* we have again three groups of dates:

A. (1) $ju = p + 29 = m + 9$[62]
(2) $\overline{ju} = \text{I } 10$[63]

---

55 Berol 84 9ᵇ I, 14—16; Princ 5884 14ᵇ III, 10—12; Scal p. 675, 10; Upps 3 70ᵇ, 16/71ᵃ, 1; Vat 119 88ᵃ I, 6—8; Vat Borg 17 2ᵇ.

56 Berol 84 13ᵃ II, 17—20; Princ 5884 14ᵇ III, 14—17.

57 Princ 5884 6ᵃ III, 15—17; Upps 3 72ᵃ, 3—5; Vat 15 279ᵃ II, 22—279ᵇ I, 2; Vat 88 50ᵇ I, 15—18; Vat 119 88ᵃ I, 8—10.

58 Esther IX, 21.—Perhaps the uncanonical date for Purim is somehow related to the fact that Esther is the only biblical book not represented (at least so far) in the Qumrān material; cf. Dupont-Sommer [1977] p. 661.

59 Cf. below s. v. Seasons, p. 206.

60 BM 815 24ᵇ II, 7/8; BM Add 24995 20ᵇ I, 21/22; EMML 215 2ᵃ II, 4/5; Vindob 6 23ᵃ II, 10 f. Cf. also Mauro da Leon. [1943] p. 307. Weld Blundell, Roy. Chron. p. 501 incorrectly identifies "Judith" with the "Fast of the Apostles" (*sh*). Cf. below p. 151.

61 Princ 5884 14ᵇ III, 17; EMML 215 7ᵃ I, 7; BM Add 24995 25ᵃ II, 3.

62 Scal p. 675, 12.; EMML 215 62ᵇ II, 27 f.

63 Upps 3 72ᵃ, 5/6; Vat 15 279ᵃ I, 17—20; Vat 61 6ᵇ (Mauro da Leon. [1943] p. 323/326, emending (p. 307) I 10 to I 4). Vat 18 253ᵃ, 12—14; day number omitted.

B. (3) $ju = p + 18 = pu + 7 = m + 28$[64]

(4) XII 13 $\leq ju \leq$ I 11[65]

(5) $\overline{ju} =$ I 4[66].

The boundaries (4) agree with the boundaries for *m* and for *pu* (above p. 145 (2)) and also (4) and (5) are parallel to (2) and (3) for *pu* and $\overline{pu}$. Similarly the relations in A are equivalent to A for *ce* (p. 143). Hence (3) of B could concern the year after (1) in A, again as with *ce*.

C. Unrelated and unexplained remain the boundaries

(6) XII 30 $\leq ju \leq$ I 28[67]

excepting for the fact that both (4) and (6) ignore the epagomenal days. This is another indication that "XII" and "I" originally did not belong to the time of the year where the epagomenal days of the Alexandrian calendar are located but to a year that started with Nisan. On the other hand, it can hardly be doubted that such an origin has long been forgotten by the Jews living in the Diaspora.

B. Christian Feasts

For the New Years Day cf. s. v. Ṭentyon and Day of John (p. 218).

1. Nineveh (n)

Also called "Fast of Nineveh"[68], commemorating Jonah's prediction of the destruction of Nineveh[69]. This feast is known in the Coptic[70] as well as in the Syrian Church[71]. Cf. also Bīrūnī, Chron. p. 311, 22—26.

---

64 EMML 215 7ᵃ I, 6—8; Scal p. 675, 11; Vat 18 252ᵇ, 13—253ᵃ, 1; Vat 61 4ᵇ (Mauro da Leon. [1943] p. 322/325).

65 EMML 215 7ᵃ I, 8—10; cf. also Mauro da Leon. [1943], p. 307.

66 Berol 994 38ᵃ—65ᵃ; BM 816 1ᵃ—5ᵃ; EMML 215 7ᵃ I, 11 f. (giving incorrectly XII 4 instead of I 4).

67 Princ 5884 14ᵇ III, 17—20 (giving incorrectly I 27 instead of I 28).

68 E. g. in EMML 215 (6ᵇ I, 9/10; I, 15 etc.); Upps 3 (72ᵇ 9; 14 etc.).

69 Rüppell, Reise, II p. 39.

70 Sobhy [1942] p. 173. Burmester [1951] p. 316/7: "This fast . . . recalls the penance of the Ninevites at the preaching of Jonah", therefore also called "Fast of Jonah."

71 Bar Šinaya (about 1230), trsl. Chabot, p. 153 Table (= Saliba [1970] pp. 202—204), trsl. Delaporte, p. 351, pp. 357—359. Cf. Weld Blundell, Roy. Chron., p. 501, n. 1.

2. Beginning of Fast (*bf*)

The Monday initiating Lent, also called "Fast of our Lord"[72]. The earliest possible date, VI 1, falls, of course, in the same years as the earliest dates for Easter, i. e. in the years $k = 491, 54, 149, 244$ ($c = 16$). One text[73] gives instead $k = 35, 130, 225, 320$ ($c = 16$), one 76-year cycle more, corresponding to the era D.

3. Mount Olive (*mo*)

According to Vat. Cat. p. 215: μεσονηστεία. Chaine, Chron., p. 173 gives VI 27 as the earliest day, instead of VI 28[74]. The consistent arithmetical structure of the dates of Christian feasts characterizes such an isolated change as an error.

4. Hosanna (*ho*)

For the origin of this term, cf. Patrol. Orient. 37 p. 52, n. 16.

5. Easter (*f*)

Cf. also s. v. Calendar, Parameters, p. 44 and s. v. Easter, p. 96. The limits for Easter are

$$\text{VII } 26 \leqq f \leqq \text{VIII } 30. \tag{1}$$

Since the 532-year tables give only day numbers, no month names, the numbers from 26 to 30 could either belong to VII or to VIII. This ambiguity is in principle avoided in the tables by the addition of a gā (usually written in red) to those numbers which belong to month VII, i. e. Magābit. But scribal errors and omissions abound in all manuscripts, and it is therefore useful to have rules which resolve such ambiguities.

We first observe that the addition of 30 days changes the weekday of a date by 2. Since f must be a Sunday, only one of two possible months can be correct. Now the ṭentyon t gives the weekday of I 1. If the interval from I 1 to Easter Sunday is $\Delta t \equiv \delta \mod 7$, we must require that

$$t + \delta \equiv 5 \mod 7 \tag{2}$$

5 being Sunday in the norm of $t = 1$ meaning Wednesday. Consequently we see that an Easter date VII 30 corresponds to $\Delta t = 209 \equiv \delta = -1$

---

[72] E. g. BM 815 25ᵇ I, 16/17.—For the interval of 55 days from *bf* to *f* cf. e. g., Rüppell, Reise, II p. 39.

[73] Princ 5884 6ᵃ III, 19—6ᵇ I, 4.

[74] Cf. above p. 133.

mod 7 which, according to (2), is only compatible with $t = 6$. In other words Easter can only be as early as VII 30 in a year with $t = 6$, i. e., when I 1 is a Monday. Similarly $f =$ VII 29 implies $t = 7$, etc., until $f =$ VII 26 when $t = 3$[75].

These remarks allow us to confirm a statement found in some texts[76] to the effect that each 532-year cycle contains 48 cases in which Easter falls into the month Magābit (VII), 484 times into Miyazyā (VIII). It obviously suffices to show that the same distribution holds for n in relation to the dates 17 to 21 in the months V and VI. Furthermore, since n is uniquely determined by t and the cycle number c, i. e. within $7 \cdot 19 = 133$ years, it suffices to show that n falls exactly 12 times into V within each 133-year cycle. Since n can vary according to

(3) $$\text{V } 17 \leqq n \leqq \text{VI } 21$$

we have to look at the distance $\Delta t \equiv \delta$ mod 7 from I 1 for the ambiguous dates 17 to 21. Hence we consider the following table

Table II

n	in V			in VI		
	Δt	δ	t	Δt	δ	t
17	136	3	3	166	5	1
18	137	4	2	167	6	7
19	138	5	1	168	0	6
20	139	6	7	169	1	5
21	140	0	6	170	2	4

in which the values of t are chosen so as to satisfy the relation $t + \delta \equiv 6$ mod 7 in order to make n a Monday. If we now look at the table which gives n as function of t and c[77] we find the following cases which involve the ambiguous dates $n = 17$ to 21:

[75] Cf. s. v. Calendar, Appendix I, p. 58.

[76] E. g. Princ 5884 5[b] I. Dillmann, Br. Mus. Cat. p. 44, No. 37, 1a, misread 48 as 38.

[77] Cf. s. v. Calendar p. 50, Table XIII.

Table III

$t =$	1	2	3	4	5	6	7	in V
$c =$ 2						21		1 case
5	19	18				21	20	4 cases
8	17			21	20	19	18	no
11	17						18	no
13						21	20	2 cases
16	19	18	17			21	20	5 cases
19	17				20	19	18	no

Comparison with Table II shows which combinations of t and n belong to month V. Their number is indeed $1+4+2+5=12$ corresponding to the cycle numbers $c = 2$, 5, 13, and 16. The same must hold for the dates of / that fall into month VII. In other words it is necessary and sufficient for an Easter date to belong to month VII that $c = 2$, 5, 13, or 16 and that the date belongs to the interval from 26 to 30[78]. These cycle numbers are the same for which Passover falls into month VII[79].

The preceding discussion also permits verification of another statement[80]: In one 532-year cycle Easter can only four times fall on the last possible date, i. e., VIII 30. Indeed, the corresponding latest date for n is VI 21 and table III shows that the date 21 appears only for $c = 2$, 5, 8, 13, and 16. In order to make n the latest date in VI we see from table II that we must have $\Delta t = 170$, hence $t = 4$. Returning to table III we find that only $c = 8$ agrees with $t = 4$, i. e., once in 133 years, thus four times in one 532-year cycle. The cycle numbers for these years in the era G are $k = 27$, 122, 217, and 312[81]. The earliest dates occur in G for $k = 491$, 54, 149, 244[82].

[78] As usual the entries in the manuscript tables are sloppy. Vat 119 lists only 21 cases, Ṭānā 34 has 43. Combined, the two texts note 46 cases. In BM 815 and BM 827 the cases with gā are numbered on the margin but reach only 44.

[79] Cf. s. v. Calendar p. 59, Table "XIX".

[80] Princ 5884 6ª III, 19—6ᵇ I, 9.

[81] The cycle numbers in Princ 5884 are 76 higher because they belong to the era D.—For the era G cf. BN Abb 37 4ᵇ I, 3—11; Marḥa 'E p. 379 I, 10—14.

[82] Listed in BM Add 16217 15ª II, 7—9 for the era D.

Another way of determining the extremal Easter dates is based on the following simple consideration[83]. The earliest Easter Sunday will be found when the earliest Passover date is a Saturday because then the very next day will be Easter. Similarly the latest date will be a week after the latest possible Passover, if this is a Sunday. Now $p = m + 190 \equiv m + 1 \bmod 7$. Hence the desired extremal dates of p will correspond to m on Friday (i. e. $t = 3$) and Saturday ($t = 4$) respectively. Hence we must look for the coincidence of $m =$ I 15 (i. e., $c = 16$) with $t = 3$ and of $m =$ II 13 (i. e., $c = 8$) with $t = 4$. The 532-year tables show that the first case takes place for $k = 54$, the second for $k = 27$, both combinations being repeated three more times at intervals of 95 years (mod 532).

6. Synod of the Apostles (*ra*)

Probably commemorating the congress of the Apostles described in Acts 15. In Vat. Catal. p. 59 rakeb is translated as "festum conventus (μεσοπεντεκοστή), p. 215 as "conventus presbyterorum". Cf. also Guidi in CSCO 25, p. 83 n. 1.

7. Ascension (*asc*)

Preceding Pentecost by 10 days, thus baʿāla 40, "Feast of 40 (days)".

8. Pentecost (*pt*)

Baʿāla 50; also called parāqlitos "Paraclet"[84]. Cf. for the Jewish predecessors of this feast, Charles, Jubilees, p. 52, n. 17; p. 106 f. note.

A "Prayer of Pentecost" (mehlā baʿāla 50) falls three days after Pentecost according to Scaliger's text[85].

9. Prayer for Salvation (*md*)

Mehlā deḇnat, not to be confused with the Jewish feast ṣoma deḇnat, "Fast of Salvation", i. e., Purim[86]. Even if the manuscripts occasionally use the wrong terminology[87], one can easily establish the association of

---

83 Described in BM Add 24995 24[a] II, 12—24.

84 Vat. Cat. p. 709, 3. Abbreviated to parā in the table Vat Borg 17 1[b] heading.

85 Scal p. 676, 2 f. Scaliger gives incorrectly behela instead of mehlā that is actually found in the text.

86 Cf. above p. 144.

87 Cf. e. g., the colophon B. M. Catal. Wright No. 183; cf. below p. 153. Also Vat. Cat. p. 709, 3 (Vat 232, 102[a]) where the date $n + 1$ establishes the feast as Christian.

a given date with either Christian or Jewish feasts by the difference in arithmetical relations[88].

10. Fast of Apostles (sh)

Two different traditions seem to be reflected in our texts: one group places the feast on Monday following Pentecost ($sh = pt + 1 = n + 119$), the other one week later ($n + 126$). Cf. Rüppell, Reise, II p. 40: beginning the first Monday after Pentecost. For the Coptic usage, cf. Sobhy [1942] p. 172 ff. and the last column of the tables, pp. 192 to 197[89]. Reference to this feast is also found in the Didascalia[90].

Weld Blundell says[91] (without citing his sources): "Fast of the Apostles, also called Judith, fast of Trinity Monday". But "Judith" is a Jewish festival[92] and therefore cannot remain on a fixed weekday.

11. Preaching of Nativity; "Prayers"

"Moveable" Christian feasts are ordinarily understood to be feasts whose dates range, parallel to Nineveh or Easter, over an interval of 35 days, their weekdays being fixed[93]. There exists, however, another group of feasts which always fall on the same weekday but whose dates vary only over 7 days. Such a feasts is the "Preaching of Nativity"[94] whose dates are made dependent on the weekday of "Elias", the first day of Taḫśāś (IV)[95], by the following rule:

if "Elias" falls on a Tuesday, then "Preaching" is on the		13th (of IV)
	Wednesday	12th
	etc.	etc.
	Monday	7th.

These intervals between "Elias" (IV 1) and "Preaching" are chosen in such a way that they always lead to a Sunday.

---

88 Cf. above p. 131 and 132.

89 With the numerical value 50–*bf*.

90 Harden, Didasc., p. 135.

91 Weld Blundell, Roy. Chron., p. 501.

92 Cf. above, p. 145.

93 Cf. above, p. 132.

94 sebkata gēnā, meaning the preaching of John the Baptist, in which he announced the coming of Christ; cf. Velat, Me'eraf p. 32, 2. Scaliger (p. 677) translated gēnā as "penitence", thus relating it to ganaya instead of to the Greek γέννα; on p. 703, however, he has "Adventus Dominici".

95 BM 734 282ᵇ II, 18—22; Bodl 21 1ᵃ II, 5—8; Vat 15 280ᵃ I, 20 to II, 11 have "Beginning of Tāḫśāś" where the other texts have "Elias" (Berol 84 10ᵇ I, 11—19; BN Abb 37 4ᵇ I, 22; Princ. 5884 6ᵃ III, 6/7; Scal p. 677, 23; Upps 3 75ᵇ, 7/8; Vat 69 132ᵃ, 1—8; Vat 88 51ᵃ I, 15).

Parallel with this feast is a "Prayer for the Preaching of Nativity" (mehlelā sebkata gēnā), preceding the former by two days. This is again formulated[96] in relation to "Elias" on IV 1:

if "Elias" (IV 1) falls on a	Wednesday, then "Prayer" on the	10th
	etc.	etc.
	Monday	5th
	Tuesday	11th.

Again the resulting weekday is alsway the same (Friday).

There are several more public prayers which depend on the weekday of some preceding civil date, so that the resulting date falls either on a Wednesday or a Friday[97], the canonical Christian Fast Days[98]. For example "Michael" on X 12 determines the dates for the "Prayer for the Beginning of the rainy season (keramet)"[99]; the weekday of the civil New Years Day (i. e. *j* or *t*) determines the "Prayer of the Cross"[100], and the first epagomenal day fixes the date for the "Prayer of John"[101].

7. *Examples*

1. Wright, Br. Mus. Cat. No. 180, p. 117 (BM 567 128ª):

በሠሉስ ፡ ሠርቀ ፡ መስከረም ፡	on Tuesday beginning of Maskaram (I)
ወንጌላዊ ፡ ዮሐንስ ።	Evangelist: John
፮ጥንቴዎን ፤	$t = 6$ (sic! error for 7)
፯አበቅቴ ፤	$e = 7$ (sic! error for 6)
፳፬ለመስከረም ፡ መጥቅ ፤	$m =$ I 24
፳፯ለጥር ፡ ነዌ ።	Nineveh: V 27

---

[96] Vat 69 134ᵇ 8—135ª 11.

[97] Expressly stated in BN Abb 37 4ᵇ II, 36/37. Rules of this type are found, e. g., in Berol 84 10ᵇ I, 10—II, 24; BN Abb 37 4ᵇ II, 6—37; Upps 3 73ᵇ, 1—74ª, 17; Vat 69 132ª, 8—133ᵇ, 13.

[98] Cf. E. Schwartz [1906], Ges. Schr. 5, p. 34.

[99] The "Beginning of keramet" itself is X 25; cf. s. v. Seasons, p. 204.

[100] The date for the "Exaltation of the Cross" is I 17; cf. Velat, Meʿeraf p. 25.

[101] These days are Wednesdays for XII 29 or 28 and for epag. 4, 3, 2, Fridays for epag. 4 and 3. Thus they precede the "Day of John", i. e. I 1.

፲፩ለየካቲት ፡ በዓተ ፡ ጾም ፤	Beginning of Fast: VI 11
፬ለሚያዝያ ፡ ፋሥሕ ።	Passover: VIII 4
፮ለሚያዝያ ፡ ፋሲካ ።	Easter: VIII 6
፲፭ለግንቦት ፡ ዕርገት ።	Ascension: IX 15
፳፭ለግንቦት ፡ በዓለ ፡ ፶ ።	Pentecost: IX 25
፫ለሰኔ ፡ ጾመ ፡ ሐዋርያት ፤	Fast of Apostles: X 3

Commentary

I 1 = Tuesday, thus $t = 7$ (not 6)
ε = John = 4, thus $G \equiv 0 \mod 4$
$e = 6$ because $e + m = 30$
m = I 24, $t = 7$ thus m = Thursday, thus $n = m + 123$; $m = 24$ thus
n = V 27 $\quad n = 24 + 123 = 147 =$ V 27
$146 \equiv 6 \mod 7$
I 1 = Tuesday, thus V 27 = Monday
for $t = 7$, $n = 27$: $c = 2, 7, 10, 18$[102]
for $G \equiv 0 \mod 4$ and $G \equiv c \mod 19$: G = 344, 368, 352, 360[103]
$bf = n + 14 = 147 + 14 = 161 =$ VI 11
$p = m + 190 = 24 + 190 = 214 =$ VIII 4
$f = n + 69 = 147 + 69 = 216 =$ VIII 6
$asc = n + 108 = 147 + 108 = 255 =$ IX 15
$pt = asc + 10 =$ IX 25
$sh = n + 126 = 147 + 126 = 273 =$ X 3

2. Wright, Br. Mus. Cat. No. 183, p. 118 (BM 570 132[b]):

በ፫፻፴፫ዓመት ፡ ምሕረት ።	in the year G 333
እሑድ ፡ ይሠርቅ ፡ መስከረም ።	Sunday began Maskaram (I)
ወንጌላዊ ፡ ማቴዎስ ።	Evangelist: Matthew
ዓበቅቴ ፡ ፴ ።	$\varepsilon = 0$

102 Cf. s. v. Calendar, Parameters, n, p. 50 Table XIII.
103 Cf. Example 2.

መጥቅዕ ፡ ፳፩ ።	$m = 21$
አሠ ፡ ፳ወ፱ ፡ ለጥር ፡ ነነዌ ።	Nineveh: V 29
አሠ ፡ ፲ወ፫ለየካቲት ፡ በዐት ፡ ጾም ።	Beginning of Fast: VI 13
አሠ ፡ ፲ለመጋቢት ፡ ደብረ ፡ ዘይት ።	Mount Olive: VII 10
አሠ፩ለሚያዝያ ፡ ሆሳዕና ።	Palm Sunday: VIII 1
አሠ ፡ ፰ለሚያዝያ ፡ ፋሲካ ።	Easter: VIII 8
አሠ ፡ ፪ለግንቦት ፡ ረክብ ።	Synod: IX 2
አሠ ፡ ፲፮ለግንቦት ፡ ዕርገት ።	Ascension: IX 16 (sic, error for 17)
አሠ ፡ ፳ወ፯ለግንቦት ፡ በዓለ ፡ ሃ ።	Pentecost: IX 27
አሠ፴ለግንቦት ፡ ጾመ ፡ ድኅነት ።	Thanks[104] for Salvation: IX 30
አሠ፭ለሰኔ ፡ ጾመ ፡ ሐዋርያት ። ።	Fast of Apostles: X 5

Commentary

G 333

I 1 = Sunday, thus $t = 5$

ε = Matthew = 1 G $333 \equiv 1 \mod 4$

$e = 9$

$m =$ (I) 21, $e + m = 30, t = 5$ thus m = Saturday thus $n = m + 128$

n = V 29 $m = 21$ thus $n = 21 + 128 = 149$ = V 29

for $t = 5, n = 29$: $c = 4, 7, 10, 15, 18$[105].

G $333 \equiv 10 \mod 19$, thus $c = 10$

$bf = n + 14 = 149 + 14 = 163$ = VI 13

$mo = n + 41 = 149 + 41 = 190$ = VII 10

$ho = n + 62 = 149 + 62 = 211$ = VIII 1

$f = ho + 7$ = VIII 8

$ra = n + 93 = 149 + 93 = 242$ = IX 2

---

104 ṣoma for mehla.

105 Cf. Example 1

$asc = n + 108 = 149 + 108 = 257 = \text{IX } 17$ (not 16)
$pt = asc + 10 = \text{IX } 27$
$md = n + 121 = 149 + 121 = 270 = \text{IX } 30$
$sh = n + 126 = 149 + 126 = 275 = \text{X } 5$

3. Dillmann, Berlin HSS, p. 73 (Berol 84 1[a]):

በ፷፻ በ፱፻ በ፲ወ፮ ዓመተ፡ ዓለም።	in the year W 6916
በ፫፻በ፶ወ፫ ዓመተ፡ ምሕረት።	in the year G 353
በወርኀ፡ ሰኔ፡	in the month Sanē (X)
እም፡ ፬ቱ፡ ለሠርቅ፡ ሌሊት፡	on the 4th in evening epoch
ወተሡዑ፡ ለሠርቅ፡ መዓልት፡	on the 9th in morning epoch
በዕለተ፡ ቀዳ ም።	on a Saturday

Commentary

The year number is to be interpreted as W 6916 + 353 = W 7269. The purpose of this formulation is to indicate a date in the 14th 532-year cycle of the World (because 6916 = 13 · 532)[106]. A similar formulation is found in Weld Blundell, Roy. Chron. p. 47, 27[107]: ba 5000 wa 500 ʿāmata ʿālam ba 1000 wa 770 wa 7 ʿāmata meḥrat, i. e. W 5500 + 1777 in order to indicate years of the era of Incarnation that begins after the completion of 5500 years of the World.

Hence our text was written in the year G_3 353 = J 1769 ~ A. D. 1777[108]. For G 353 = $c \equiv 11 \bmod 19$, one finds in the 532-year tables $t = 6$, $e = 20$. Hence one finds for the lunar date that corresponds to the civil date X 9[109]

$$\bar{d} = 1 + e + [\mu/2] + d = 1 + 20 + [9/2] + 9 = 30 + 4 \equiv 4 \mod 30$$

in agreement with the text.

Finally: since $t = 6$ = Monday and X 9 = I 1 + 278 $\equiv$ I 1 + 5 mod 7 we have indeed for X 9 a Saturday.

---

106 On fol. 18[b] II, 3 it is correctly stated that the last year of the 13th cycle of the era W corresponds to the year D 1140. Indeed D 1140 = W 6916.

107 The translations on p. 269 and note 1 are wrong.

108 Dillmann, p. 73; "geschrieben im Jahre 1777, aber sehr incorrect".

109 Cf. above p. 141 (5).

Feria

It is convenient to use the medieval term "feria" for the numbering of weekdays, beginning with Sunday = 1. No other counting of weekdays is here referred to by this term[1].

---

[1] Neither Ge'ez nor Greek knows this terminology.

Fixed Stars

See Stars (p. 216).

Gates

See also: Windows (p. 227).

One of the most characteristic concepts of the astronomy traditionally connected with the name of Enoch is the description of risings and settings of sun and moon through six "gates" (ḫoḫet), symmetrically arranged at the eastern and western horizon[1]. In the course of one year the rising- (and setting-)point of the sun moves over an extended arc of the horizon which amounts very nearly to 30° to the south and to the north of the east- (and west-)point of the horizon, assuming geographical latitudes between 31° (Alexandria) and 38° (Athens)[2]. Hence the rising (and setting) amplitude of the sun can be assumed to be about 60°. The six equal arcs, of 10° each, from 30° south of East to 30° north of it (and similarly in the West) can be shown to correspond reasonably well to the six "gates", numbered from south to north. In other words the rising from a "gate" refers to a section of the horizon where the sun or the moon appears.

---

[1] These concepts existed already in the Aramaic version of the astronomical Book of Enoch; cf. Milik, Enoch, p. 203.

[2] For the latitude $\varphi = 36°$ (Rhodes) this amplitude is almost exactly 30°. In ancient astronomy this latitude is frequently chosen as representing a mean value for geographical conditions. Cf. for all numerical data the discussion in Neugebauer [1964].

II.

CATALOGUE OF TOPICS

Abbreviations of Technical Terms

In the column headings of tables one often finds abbreviations which simply consist of the initial syllables of the complete words (e. g. ʿāw qama for ʿāwda qamar[1]). Similar abbreviations occur in rules which repeat indentical phrases many times[2]. There exist, however, also abbreviations which are so short or which consist of internal or final syllables that their meaning can only become clear through the understanding of their function. Such abbreviations are nowhere listed in the dictionaries or in the literature and thus can become a major obstacle for the interpretation of a text or a table. It is for this reason that I give here a list of such non-trivial abbreviations which I encountered in my material.

ʾa	ʾatārd	Mercury (Arabic)
ʿā	ʿāmat	year
ʿā	saʿāt	hour
ʿā ḍa	ʿāwda ḍaḥay	solar cycle (28 years)
bo	ʾalbo	zero
da	daḫārāy	second (month)
da ḍa	ʿāwda ḍaḥay	solar cycle (28 years)
ḍu	ḥeḍuḍ	hollow month (29 days)
ʿe	ʿelatātihu	its days
fāzagā	fāsikā za magābit	Easter falls in month VII
gā	magābit	month VII
gā tā		cf. gā, cf. tā
ʾi	ʾiyobēl	jubilee (49 years)
lā qa	lāʿelāy qamar	latest date for Easter (i. e. VIII 30)
lafā	lafāsikā	for Easter
lafāzagā	lafāsikā za magābit	of Easter falling in month VII
lami	lamiyāzyā	of month VIII

[1] Cf., e. g., Rhodokanakis [1906] p. 82; Vat. Catal. I p. 484; Hammerschmidt, Ṭānā I p. 157 f.

[2] E. g. BM Add 16217 14ᵇ: baʿā(ta) ṣo(m) la(ʾema) ra(kabka).

This process continues as is illustrated in the following table[6].

"Hebrew"	"Roman"	days	In Ethiop. calend. first day	last day	"parts" day	night	Sunrise in gate
Nisan (VII)	April	30	VIII 6	IX 5	10	8	4
Ayar (VIII)	May	31	IX 6	X 6	11	7	5
Ḥaziran (IX)	June	30	X 7	XI 6	12	6	6
Tamuz (X)	July	31	XI 7	XII 7	11	7	6
Ab (XI)	August	31	XII 8	I 3	10	8	5
Elul (XII)	Sept.	30	I 4	II 3	9	9	4
Tišre I (I)	Okt.	31	II 4	III 4	8	10	3
Tišre II (II)	Nov.	30	III 5	IV 4	7	11	2
Kanun I (III)	Dec.	31	IV 5	V 5	6	12	1
Kanun II (IV)	Jan.	31	V 6	VI 6	7	11	1
Šebath (V)	Febr.	28	VI 7	VII 4	8	10	2
Adar (VI)	March	31	VII 5	VIII 5	9	9	3

The first column gives the months of the Syriac calendar (ascribed by the text to the "Hebrews"). In the next column these months are simply identified with the months of the Roman calendar whose lengths are given in the third column, similar to (2). The next two columns give the dates in the Ethiopic calendar of the first and last day of each of the Roman months[7]. The remaining columns establish the connection with the Enoch material. The length of daylight varies linearly between 6 and 12 "parts" and the sun is assumed to rise each month in another gate, except for doubling in the solstitial gates 1 and 6. No doubt we see here the process of transition from the original concept of "gates"

[6] From Vindob 6 1ᵃ, 1 to 2ᵃ, 21; 3ᵃ, 10 to 15.

[7] The epagomenal days are correctly accounted for between XII and I 1 = August 29.

of sunrise and sunset to a new type of 12 "gates" which are the domiciles of the sun during the 12 months of the year and thus tend to become the equivalent of the 12 zodiacal signs. Consequently each gate measures 30 "stades"[8], i. e., 30 degrees.

Returning to the original concept of the "gates" as six equal arcs of about 10° in length[9] with the east-point as center, we can now discuss the pattern which associates the risings of the moon with the gates. We have several texts[10] that list the number of days for which the moon rises (or sets) in each gate during a sequence of 12 months, alternatingly full and hollow. In the texts these months are usually named in the order of the Ethiopic months, a procedure which in fact is incompatible with a real lunar year. Consequently we are counting here the months from "1" to "12" (beginning at the vernal equinox). Actually we do not know how the months were named by the inventors of this arithmetical pattern, though Nisan seems likely for the first month. In the Book of Enoch, at the description of the variation of the length of daylight and night[11], the months are simply numbered from 1 to 12, beginning with gate 4 after the vernal equinox.

In Table I we give a restoration of the whole scheme. It deviates from the numbers found in the texts only in a few cases which are either trivial copyist errors or desperate attempts to reconcile the pattern at its end with a year of 364 (or 365) days instead of the 354 days of the original scheme. The distribution of the intervals of length 1 or 2 days has become rather arbitrary as long as the proper totals 29 or 30 are upheld.

Two facts can be noted immediately: first, the scheme is periodic, i. e. the pattern of month 12 leads again to the pattern of month 1; secondly, in each month the moon starts and ends in the same gate. For example, in month 1 the moon moves from gate 4 to gate 6, then back to gate 1 and forward again to gate 4. Consequently the next month begins in gate 5 as is necessary because the sun rises now in gate 5 after

8 Bonwetsch [1922] p. 68 (ch. XIII); Vaillant, Secrets, p. 13.

9 It is only for modern convenience that we measure these arcs in degrees. The Book of Enoch knows, of course, no such unit (misunderstood: Milik, Enoch, p. 14).

10 Grébaut [1918, 2] (from BN 64); also Berol 84 30^{a} II, 10 to 31^{a} III, 2; Vat 106 91^{b} I, 7 to 92^{b} I, 21; Vat 119 66^{a} I, 1 to 69^{a} I, 10. Partial English rendering: Milik, Enoch, p. 275 f.

11 Enoch, ch. 72; cf. s. v. Length of Daylight, p. 180.

Table I

Months	1	2	3	4	5	6	7	8	9	10	11	12	1	Months
Gates														Gates
4	2													4
5	2	2												5
6	8	8	4	4										6
5	2	2	2	2	2									5
4	1	1	2	2	1	2								4
3	1	1	1	1	1	1	2							3
2	2	2	2	2	2	2	2	2						2
1	8	7	8	7	8	7	8	7	4	4				1
2	2	2	2	2	2	2	2	2	2	2	2			2
3	1	1	1	1	1	1	1	1	2	2	1	2		3
4	1	1	2	2	1	1	1	1	1	1	1	1	2	4
5		2	2	2	2	2	2	2	2	2	2	2	2	5
6			4	4	8	8	8	8	8	7	8	8	8	6
5					2	2	2	2	2	2	2	2	2	5
4						1	1	1	1	1	1	1	1	4
3							1	2	2	2	1	1	1	3
2								1	2	2	2	2	2	2
1									4	4	8	7	8	1
2											2	2	2	2
3												1	1	3
4													1	4
Days	30	29	30	29	30	29	30	29	30	29	30	29	30	Days

it rose in the preceding month in gate 4. Only at the solstices does the sun rise in the same gate during two months (months 3/4 and 9/10). At first sight the change between a stay of the moon for 8 days and for 1 day in one gate seems strange. In fact these intervals are of the right order of magnitude as can be seen by computing the time intervals for which the ecliptic intersects the horizon within the same gate[12].

For the sun the distribution of time should be even more unequal; the times in the gates 4 to 6 and back into 4 (and similarly for 1 to 3) should be roughly as follows:

Gate 4	20^d	20^d
5	24	24
6	47	47

We have seen in (1), p. 157, what scheme has been adopted instead. Finally, when the original meaning of the gates was no longer understood, the times for the sun in each quadrant were simply taken to be 30, 30, 31^d[13]. It is characteristic for the uneven level which is preserved in the "Book of Enoch" that this scheme is already found in the main astronomical section[14].

Also the descriptions of the physical structure of the gates are far from consistent. In Enoch ch. 36, 2 f., e. g., we are told that there are "small gates" above the three eastern gates and that the stars come forth from these small gates. On the other hand there are 12 "windows" above gate 4 in the east, ejecting flames[15], and "many windows" to the right and to the left of each gate[16].

It is not surprising that in this chaotic mixture of contradictory fantasies the original meaning of the gates was lost and "some" could say[17] that the moon rises during 8 days in the outermost gates (1 and 6), 7 days in the central gates (3 and 4), but never in gates 2 and 5—which indeed would be in excess of the 30 days attributed to the other four gates. The numbers 8 and 7 reflect obviously the numbers in Table I for the gates 1 and 6 and were incorrectly distributed. When one finally again finds data which make sense astronomically, it is in connection with Arabic terminology[18] and based on zodiacal signs[19].

---

12 Cf. for details Neugebauer [1964] pp. 55—57.

13 E. g., Berol 84 27^a III, 12 to 28^a III, 13.

14 Ch. 72, 6 to 37.

15 Enoch, ch. 72, 7.

16 Enoch, ch. 72, 3.

17 Berol 84 31^b III, 22 to 32^a II, 3.

18 E. g. Vat 54 88^a to 89^a or Ambros 22 182^a, 2 to 182^b, 11 and 1^a, 1 to 2^a, 2 (incomplete).

19 Conversely, in late antiquity, zodiacal signs were called "gates of the sun", e. g. Macrobius (around A. D. 400) says that Cancer and Capricorn "portae solis vocantur" (Saturnalia, I, 17, 63).

Gematria

The interpretation of words, in particular of proper names, as numbers is a widespread practice in the hellenistic world[1]. Since this procedure is based on the existence of alphabetic numerals its Greek origin (and later transmission to Hebrew and Arabic) seems evident[2]. The Ethiopic numerals, however, are not letters of the Ethiopic script[3] and therefore all kinds of tricks had to be invented for the transformation of words into numbers[4].

Some forms of gematria are well attested in Ethiopic texts. For example, the numerical equivalents of a person's name and of the names of friends and relations are added and the total is reduced modulo some astrological parameter, e. g. mod 28 in the "Computus of Stars and Lunar Mansions"[5]. The remainder (between 1 and 28) then points to a corresponding lunar mansion and its specific qualities. Similarly a reduction modulo 12 leads to a zodiacal sign[6]. Obviously all this is due to Arabic influence.

The technical term for this arithmetical operation is gedef, "cast out", (namely multiples of the modul)[7] but it has been almost invariably mistranslated (e. g. "diminue de 7", or "retranche 2", or even "Alternative")[8]. Exactly the same terminology is commonplace in calendaric computations for reductions modulo 19, 30, or 7, etc.

A peculiar form of gematria is connected with a number called "benediction" (barakat) and with the epact (e) of a year. The number barakat (β) is found from

$$\beta \equiv e + 4 \mod 7,$$

---

1 Cf. e. g., Dornseiff, AMM, Stoicheia 7 (1922). Dignified classical handbooks (e. g. Pauly-Wissowa or The Oxford Classical Dictionary) do not even mention the word gematria. Cf. also Neugebauer, The Study of Wretched Subjects, Isis 42 (1951), p. 111.

2 Attempts to demonstrate Babylonian antecedents (Dornseiff, AMM, p. 91) can be disregarded because they were based on speculations about Babylonian arithmetical and metrological tables before the decipherment of Babylonian mathematics.

3 They are adapted from the Greek alphabetic numerals.

4 E. g. by using the lexicographical order of the Ethiopic letters. Cf. Griaule [1934] p. 25 f. Extensive lists of correspondences in EMML 215 52[b] II, 17—53[a], II, 5; 55[a] II, 6—55[b] II, 28; 56[a], I, 21—II, 18.

5 Vat 119 fol. 42 to 49, Vat. Catal. p. 481.

6 Vat 119, Vat. Catal. p. 485, 11.

7 Cf. also s. v. Astrology, p. 20.

8 Another misleading rendering is "horoscopia 16" (Vat. Catal., p. 486, 13) for something that is nothing but a reduction mod 16.

or similar relations, and apparently represents a weekday[9]. The determination of Jacob's birthday may serve as an example[10]: add the (numerical values of the names of the) 21 Patriarchs and of Jacob's father and mother to $7e + 2\beta$; reduce the total modulo $3 \cdot 7$ (= 21); the remainder is 11; thus Jacob's birthday is the 11th of Ṭer (V).

This is only one of seven similar computations in the "Computus of the Sun, that is to say of the Dates", all of which add linear combinations of β and e to names; the modules are $1 \cdot 7$, $2 \cdot 7$, and $3 \cdot 7$[11]. The numerical equivalents for the names escape us, however.

[9] Cf. the computation of β below s. v. Magic p. 185.

[10] Patrol. Or. 6, p. 440.

[11] Patrol. Or. 6, Livre des Mystères, pp. 428—441. Grébaut completely missed the underlying arithmetical structure.

Golden Number

Scaliger[1] and others, following him[2], incorrectly identified the term maṭqeʿe (m)[3] with the "Golden Number" of the medieval computists[4]. The factual equivalent of the Golden Number is the cycle number c (from 1 to 19) in the 532-year tables.

[1] Scaliger, De em. temp., p. 690 (followed by Ludolf, according to Weld Blundell, Roy. Chron., p. 499); but correct boundaries on p. 703.

[2] E. g., Conti Rossini CSCO 58, p. 43, 14; 44, 19, etc.; Ginzel, Hdb. III p. 326.

[3] For the real meaning of this term, cf. s. v. Maṭqeʿe and Epact.

[4] Correct interpretation in Weld Blundell, Roy. Chron., p. 499; Chaine, Chron., p. 104, p. 106; Mauro da Leonessa [1943] p. 304 f., p. 305, n. 1.

Greeks

Greek influence on Ethiopia is evident from the early days of Axum[1] and remained effective, if only indirectly, through connections with the monophysite church in Egypt.

[1] Cf., e. g., Ullendorff, Ethiop., p. 54 f., about the Greek inscriptions of King Ezana (4th cent. A. D.); cf. also s. v. Conversion (p. 69). Note that the numerals of the Ethiopic script are derived from the (capital) letters of the Greek alphabet.

For the present study, however, the problems involved are much more specific. We are dealing here with two sharply distinct types of "astronomical" or calendaric material: the very primitive concepts which are more or less connected with the name of "Enoch", and the sophisticated "computus" that culminates in the 532-year tables. The Enoch-material is ultimately based on Palestinian sources, now also known from the Qumrān fragments, whereas the computus with its accurately defined calendaric rules and concepts goes back to the comparatively early Alexandrian Christianity that adapted elementary concepts of Jewish time-reckoning to the requirements of Christian liturgy.

To avoid misunderstandings I should like to stress the point that I do not see any influence of Greek scientific astronomy in the Ethiopic calendaric cycles. The astronomy of Hipparchus or Ptolemy is totally irrelevant for these tables. Nevertheless the construction of the Ethiopic tables requires a technical skill in dealing with arithmetical and astronomical material, however elementary, that considerably exceeds the competence of many a modern writer on chronology.

The agreement of the data preserved in the Easter messages of the time of Athanasius (Diocletian 44 to 89) with the 532-year tables[2] shows that these tables existed in the fourth century—and probably not much earlier in view of what we know of the gradual development of Christian customs concerning Easter. Having thus obtained a fairly secure estimate for the time of origin of the computus does not mean that we know anything about the time of transmission into Ethiopia. Only as a terminus post quem can the Era of Grace be considered, which starts 76 years later than the Era of Diocletian. Hence the period around A. D. 400 seems to be the earliest possible date, while the actual transmission into Ethiopia may have taken place considerably later, perhaps motivated by the increasing isolation from Alexandria after the Islamic conquest of Egypt.

References to the Greeks are not rare in Ethiopic treatises and tables[3]. Also "Hermes" as author[4] reflects hellenistic tradition.

---

[2] Cf. s. v. Easter, App. I, p. 99.

[3] E. g. Vindob. 6 19[b] and Princ 5884 9[a] I in the heading of 19-year tables (cf. s. v. Calendar, p. 35, n. 24). Frankf 9, colophon (Goldschm., p. 23) refers to the era of the World as "according to the Copts and the Ionians".

[4] Vindob 6 18[b]: "according to Hermes" (zaḫermēn) in the heading of the rearranged 19-year tables (cf. s. v. Calendar, p. 36). Conti Rossini [1915] p. 217, No. 238 (= BN Abb 109) in an astrological treatise, translated from the Arabic: "saying of Hermes" (misread as ḥermos and translated as Hormuz by Chaine, Abbad., p. 68, No. 109).

Shadow tables, Greek as well as Ethiopic, mention a "King Philip", in reality most likely Philip of Opus (4th cent. B. C.), referred to also in Ptolemy's "Phaseis". Synesius is associated with him in the Ethiopic tables, and is certainly Synesius of Cyrene (around A. D. 400), the pupil of Hypatia[5].

Greek influence on the Ethiopic computus is also visible in the use of loan-words for important technical terms:

'abaqtē from ἐπακτή, passim
baradyos or baryodes from περίοδος, e. g. for the 532-year cycle[6]
kēkros and kentros: cf. below, p. 175 ff.
paguemēn from ἐπαγόμεναι, passim
ṭentyon from τῶν θεῶν; cf. below, p. 219.

---

[5] Vat. 128 104ᵇ I, 3—5: "investigated by Synesius and Philip, kings of the Greeks". Cf. also Neugebauer, HAMA II, p. 872; p. 739, p. 743. The name Philip (filepos) was familiar to the scribes. Synesius, however, caused trouble. Thus we have the spellings:

sunesis	(EMML 215 72ᵃ II, 1)
sonesis	(Berol 84 15ᵇ I, 4), sonesisā (EMML 215 66ᵃ I, 25)
zenusis	(BN 50 3ᵇ marg. 3)
zanaśis	(Vat 128 104ᵇ I, 3)
zanaśiśa	(Vat 119 56ᵃ I, 2; incorrectly emended in Vat. Cat. p. 483).

[6] For 13 cycles, i. e. 6916 years: Berol 84 18ᵇ I, 1; EMML 215 64ᵃ II, 24.

"Guide for the Blind" (marḥa 'ewur)

A treatise under this title (in 10 sections) was compiled by a dabtarā (a low-ranking cleric)[1] in 1896 and sent to Guidi who published it in the same year in the Rendiconti of the Accad. dei Lincei[2], but without giving a translation or commentary.

References to the "Guide" are also found in our material[3] but the text published by Guidi (written almost completely in Amharic) is undoubtedly a composition made by the dabtarā who assembled in it what he found of interest in the Computus and in biblical chronology. Most of his material can be easily paralelled with similar sections in other treatises and in better arrangement. Hence I see no reason for accepting this

---

[1] Cf. e. g., Lifchitz, TMR, p. 9.
[2] Guidi [1896].
[3] E. g. Vindob 6 3ᵃ, upper margin: marḥa 'ewrān.

material as genuinely belonging to the "Guide". An exception is a reference to an accurate parameter (from Arabic lunar theory) which belongs to a larger group of such numbers which are also in other texts[4] assigned to the "Guide for the Blind". To mention one such parameter out of context is, of course, pointless but it characterizes the working method of our dabtarā. Also in other cases I have some doubt that he understood what he had collected. For example, the rules in Sect. 7 (p. 375 I, 25—II, 1) for the transformations from maṭqeʿe (*m*) to Nineveh (*n*)[5] are separated from the corresponding data in 375 II, 6—16 and are intelligible only if one knows beforehand what he is talking about.

Given this situation I also doubt the assertion made in the preamble of the text[6] that the "Guide" was also called "Ocean of Computus" (bāḥra ḥasāb), an identification that I have not seen anywhere else. Guidi[7] seems to assume that the "Guide" was the work of the Patriarch Demetrius[8] (written in Coptic) and transposed into three versions ("completa — media — abbreviata") in Geʿez from which our "trattatello amarico" derived. It seems to me that this genealogy does not explain the only secure element about the "Guide", i. e. the outspoken Arabic component in its contents.

A different tradition is reported by Mauro da Leonessa[9] according to which the bāḥra ḥasāb covers the "computo ecclesiastico" (movable feasts), while the "Guide" corresponds to "calendario". Admittedly this distinction is not very strictly observed; I cannot find any support in the texts for any of these classifications.

---

[4] Vindob 6 3[a] 1—9.

[5] Cf. s. v. Feasts, p. 135.

[6] P. 364 I, 3/4.

[7] Guidi [1896] p. 363.

[8] Guidi (p. 363, n. 1) refers to Renaudot, Historia patriarcharum Alexandrinorum (Paris 1713) for Demetrius's authorship of the "Guide". Actually the passage in question (read p. 25 instead of 27) contains nothing but a reference to Eutychius for the well known Ethiopic story about Demetrius' authorship ("à Spiritu sancto accepisse legitur") of the epact- and Easter Computus; cf. above p. 93.

[9] Diss. Cron. p. 55 f.

Ḥasāb

See Computus (p. 68).

Hemerology

There exists a "Computus of Lucky and Unlucky (Days)" which consists of a list of dates, one in each month, that separates the first, lucky, part of the month from the second part that consists of unlucky days. I see no rational pattern in the variation of these dates[1].

Text: ROC 18 (1913) p. 97; parallel (with small variants) Vindob 6 **31^{b}** I, 23—II, 13.

[1] What Grébaut calls a "opuscule grec analogue" (described by Nau, ROC 12 [1907] pp. 14—21) is based on individual dates, not on a two-division of each month.

Hermes

See Greeks (p. 164).

Hippolytus

See Cycles (p. 85).

Hours

Two types of "hours" must be distinguished: either the total of daylight and night, the "nychthemeron", is given a fixed number of hours of equal length, or daylight alone and night alone are divided into the same number of hours. The hours of the first type are called "equinoctial hours", in contrast to the "seasonal hours" of the second type. In terms of equinoctial hours the length of daylight varies with the seasons of the year, while the number of seasonal hours for daylight or night remains always the same. Both types of hours occur in Ethiopic texts but without differentiation in terminology (saʿātāt in in both cases).

Equinoctial hours fall again into two distinct groups: in one group, of hellenistic origin, the nychthemeron is divided into 24 "hours". The assumption of a linear variation of the length of daylight, increasing and decreasing in six steps each with the simple increment of 1 hour, leads to the extrema $M = 12 + 3 = 15^{h}$, and $m = 12 - 3 = 9^{h}$, i. e. to the ratio $M : m = 15 : 9$, which was considered, in early hellenistic times, as typical for "Greece"[1]. The other group, represented by the Book of Enoch[2], is based on the ratio $M : m = 2 : 1$. Again assuming linear variation and a difference of 1 requires a division of the nychthemeron into 18 "parts" (kefel). Indeed $9 + 3 = 12^{p} = M$, and $9 - 3 = 6^{p} = m$. In the "Secrets of Enoch" these parts are called "great hours"[3], probably in contrast to the shorter hellenistic hours.

The terminology used in Ethiopic treatises makes it sometimes difficult to distinguish between the two groups of equinoctial hours. Not only are "parts" (p) also used for sub-units of the hellenistic equinoctial hours (cf. (1 b) and (2 a)) but also the term "kentros" (kn) appears with both types of equinoctial hours (cf. (1 c) and (2 b)), denoting 1/30. In this way we have for the hellenistic hours[4]:

(1 a) $1^{d} = 24^{h}$

(1 b) $1^{d} = 360^{p}$ thus $1^{p} = 0;4^{h}$

(1 c) $1^{h} = 30^{kn}$ thus $1^{kn} = 0;2^{h}$ and $1^{d} = 720^{kn}$.

Here the "parts" represent "time degrees", commonly used in Greek astronomy[5].

An interval of two hours is occasionally[5a] denoted as one "time" (gizē) and related to the sexagesimal parts kēkros (kk) by

(1 d) $1 \text{ time} = 2^{h} = 10^{kk}$.

I do not know where this concept of "double hour" originated. A relationship to the Babylonian

(1 e) $1 \text{ uš} = 0;4^{h}$ $1 \text{ bēru} = 2^{h}$

seems to me unlikely.

---

[1] Cf. Neugebauer, HAMA II, p. 581. At a much later period $M : m = 15 : 9$ characterizes the Hellespont or Byzantium, the ratio being determined by spherical trigonometry and no longer by linear schemes.

[2] Book of Enoch, ch. 72, 10 to 32; also Vindob 6 1ª, 8 to 2ª, 21; 3ª, 10 to 15. Cf. also s. v. Length of Daylight, p. 182 n. 11.

[3] Slavonic Enoch, later version, ch. XIV (Bonwetsch [1922] p. 13); Vaillant, Secrets, p. 93. There the night is given 7 great hours of total darkness and 1 hour for what is obviously twilight; thus $1 + 7 + 1 = 9$.

[4] BM 816 18ª I, 15 to 25; II, 17 to 20. Also Berol 84 28ᵇ II, 10 to 29ª I, 1; Vindob 6 32ᵇ II, 16 to 33ª I, 16 and Berol 84 4ᵇ I, 14 to II, 3.

[5] For a different terminology, representing time degrees, cf. s. v. Kēkros, p. 177.

[5a] E.g. in a (slightly garbled) section on the linearly increasing time intervals between sunset and moonset, given in BN 160 43ª I, 3—43ᵇ I, 16.

A second group of units concerns the division of the day into 18 parts, following the Enoch tradition[6]. In the Book of Enoch itself, however, such smaller units are nowhere mentioned and it seems likely that the following relations (2 b) were only introduced as parallels to (1 c):

(2 a) $1^{d} = 18^{p}$

(2 b) $1^{p} = 30^{kn}$ thus $1^{kn} = 0;2^{p}$ and $1^{d} = 540^{kn}$.

In the Aramaic manuscripts from Qumrān[7] some early evidence seems to exist for the 18-division of the nychthemeron. Unfortunately the text is incomplete and badly preserved, so that our conclusions are not much more than suggestions of possible interpretations[8]. In the text we find mentioned days 1, 2, 3 with fractions 1/10, 1/15, 1/30, respectively, of 1/6. If we interpret 1/6 as a fraction of the shortest night (which is 6 "parts" long), the sun rising in the outermost gate (6, not 1), then we can say that we have fractions 0;6, 0;4, 0;2 of 1^{p}, i. e. a daily change of $0;2^{p}$ which corresponds to 1^{kn} according to (2 b). There is also mentioned (twice) a fraction 1/10 of 1/9 and 1/3 of 1/9. Assuming the situation at the equinoxes, the 1/9 of 9^{p} again represents 1^{p} and the fractions are $0;6^{p}$ and $0;20^{p}$. If, e. g., the $0;6^{p}$ belong to a decreasing sequence, $0;20^{p}$ to the next increasing sequence, the interval would be 14 days. In the text the $0;20^{p}$ seems to be associated with a day 15; thus $0;6^{p}$ would belong to day 1—the same as before in the shortest night. This is about as far as one can safely go; when a "star" is mentioned, one could think of some schematic description of time intervals between risings of a star and sunrise.

Seasonal hours, commonly used in hellenistic civil time reckoning, are also found in Greek and Ethiopic shadow tables[9]. This fact, evident in the 12-division of the daytime at all times of the year, agrees with the well attested hellenistic origin of these tables.

Unknown outside Ethiopic shadow tables are the names for all the hours of daylight found in a few texts. The best version is represented by Ṭānā 83 which reveals the following pattern[10]:

---

[6] BM 816 18ᵃ II, 21 to 19ᵃ I, 4.

[7] Milik, Enoch, p. 296 f.

[8] It does not make sense when Milik speaks about "movements of the stars on the celestial sphere divided into 360 degrees" (!) or of "reckoning of the astral evolutions".

[9] Cf. Neugebauer, HAMA II, p. 736 ff. Also s. v. Shadow Tables p. 209.

[10] Hammerschmidt, Ṭānā II, p. 103. The variants in Upps 42 95ᵃ and Vat 123 66ᵇ are in great disorder.

morning	third (hour)	noon	(9th?) hour	
1: high 2: low	3: high 4: middle 5: low	6: high 7: middle 8: low	9: high 10: middle 11: low	12: vesper

For the hours 9 to 11 one would expect, in analogy to the hours 3 to 5, the name "ninth (hour)". Perhaps "hour" (saʿāt) is a mistake for "ninth" (tāseʿet).

Mythological speculations did not stop at the hours. We read, e. g., about "hours of the Lord" of 83 years[11]. This number is based on the saying that 1000 years are like one day before the Lord. Since 12 of these hours suffice to give 996 $\sim$ 1000 years, we see that the "day" of the Lord covers only daytime.

Some passages, always in connection with attempts to explain the existence of a sixth epagomenal day[12], seem to imply "hours" of twice the ordinary length. For example[13], we read about "the hours of the 365 days of the year and the 3 hours which accumulate to one day in four years; thus the number of hours in one year amounts to 8763" ($= 365 \cdot 24 + 3$). Obviously we have here a flagrant inconsistency in the concept "hour". Its origin probably lies in some statement that for daytime as well as for night the yearly increment amounts to 3 hours.

---

11 E. g., Berol 84 17^a II, 22—17^b I, 13.
12 E. g., Berol 994 68^a I, 8—II, 5; Princ 5884 14^b II, 7—III, 8.
13 Vindob 16 4^b I, 5—10.

Indictio

See Eras (p. 120).

Intercalation

See Epagomenal Days (p. 113); see Feasts (p. 137); see Leap Years (p. 178).

Jesus

The era of Incarnation (J) is related to the era of the World (W), or "since Adam", by

(1) $$J\ 0 = W\ 5500.$$

The most commonly used biographical data (with their corresponding calendaric parameters as listed in the 532-year tables of the era G)[1] are shown in the following table[2]:

	W	J	G	*c*	*e*	*j*	*t*	*p*	Date
Conception	5500	0	180	9	28	3	7	VIII 12	VII 29 Sunday
Birth	5501	1	181	10	9	4	1	VIII 1	IV 29 Tuesday
Baptism	5531	31	211	2	11	6	3	VII 29	V 11 Tuesday
Crucifixion	5534	34	214	5	14	3	7	VII 26	VII 27 Friday

Consequently the Sunday of Resurrection has the date

(2) $$f = \text{VII } 29.$$

This date occurs 12 times in each 532-year cycle, at intervals of 95 years[3], three times for each Evangelist. In the era G these years are[4]

	35 Luke,	130 Mark,	225 Matthew,	320 John
(3)	119	214	309	404
	203	298	393	488

The date which is assumed for the Conception

J 0 VII 29 Sunday

precedes the date of Birth by 9 months and 5 days[5].

The above quoted dates are not the only ones attested in our texts. For example one finds the following years listed[6]

	Birth:	W 5536	thus	$k = 216$
(5a)	Baptism:	W 5566		246
	Resurrection:	W 5569		249.

[1] E. g. in Vindob 6 8[b], 9[b] at $k = 181, 211, 214$; similarly in EMML 215 87[b], 88[a].

[2] Frequently in treatises; e. g., Berol 84 39[b] I, 29—II, 6; BN 160 81[b] I, 15—82[a] II, 11; Marḥa 'E 380 I, 2—18.

[3] Cf. for this interval s. v. Ṭentyon, Appendix (p. 223).

[4] Listed in BN Abb 37 4[b] I, 11—21.

[5] EMML 215 46[b] II, 29/30.

[6] Berol 84 7[b] I, 12—17; Vat 61 (Mauro da Leon. [1943] p. 326; misreading 217 for 216? but also BM Add 19658 58[a], 8 has W 5537).

Or in greater detail[7]:

(5b)	Conception:	W 5535	VII	29		
	Birth:	W 5536	IV	29	$t = 6$	$e = 6$
	Baptism:	W 5566	V	11	$t = 1$	$e = 7$
	Resurrection:	W 5569	VIII	1	$t = 5$	$e = 11$

The values listed for t in (5b) are actually values of j. Consequently data derived from these parameters are incorrect (e. g. Easter at VIII 1 instead of VIII 4).

Even more obscure is a total of 5593 years from the time from the "first crime" (in Paradise) to the Birth of Christ, 30 more years to Baptism, and 4 (!) years to the Resurrection[8]. It is difficult to understand how such contradictory dates could be listed in our documents[9].

---

[7] BM Add 24995 32ᵃ II, 3—32ᵇ I, 7. In the Western church Baptism (on V 11) became Epiphany (on Jan. 6), since, in ordinary years, V 11 = Jan. 6.

[8] Vat 1, Catal. p. 8 (199ᵇ).

[9] For still another chronology (which assumes for the birth of Christ the year 5463) cf. above s.v. Eras, p. 126.

John, Day of

See Ṭentyon (p. 218).

Jubilees, Weeks of Ezra and Enoch

According to biblical tradition one "Jubilee" is a time interval of 49 years[1]. In our texts this interval is combined with two more similar units:

the "Week of Ezra" of 7 years[2], often simply called "Week"[3]
the "Week of Enoch" of 700 years[4].

Finally, the calendaric 532-year cycle also becomes involved in order to accumulate a total of 7000 years, the duration of the world and the date

---

[1] Levit. 25, 8. Cf. also Charles, Jubilees, Introd., p. LXVII.

[2] Grébaut [1920] p. 218, et passim; Upps 3 58ᵇ, 12 to 59ᵃ, 16.

[3] BM Add 24995 31ᵇ I, 4—11; Upps 3 58ᵇ, 1. Passim in the Book of Jubilees.

[4] Grébaut [1920] p. 219, et passim; Berol 84 22ᵃ I, 28 to II, 5; Upps 3 60ᵃ, 16 to 60ᵇ, 9.

of the Second Coming of Christ[5]. These 7000 years are "7 days of the Lord that are 7000 years for us, i. e. 2548000 days"[6]. This last number is the product of 7000 and 364, the number of days in the year of Enoch.

The shorter intervals are interrelated in many ways. The cycle of 532 years can be obtained, e. g., by the following combinations:

$$1 \text{ cycle} = 76 \text{ weeks } (= 76 \cdot 7) = 10 \text{ Jubilees} + 6 \text{ weeks } (= 490 + 42) = 532 \text{ years}$$[7]

and the Week of Enoch from

$$14 \text{ Jubilees} + 2 \text{ weeks } (= 14 \cdot 49 + 2 \cdot 7) = 700 \text{ years}$$[8].

Among the many ways of obtaining a total 7000 may be mentioned

$$\begin{aligned} 7000 &= 10 \text{ weeks of Enoch}^{9} \\ &= 141 \text{ Jubilees} + 13 \text{ weeks } (= 6909 + 91)^{10}; \\ 141 \text{ Jubilees} + 1 \text{ week } (= 6909 + 7) &= 13 \cdot 532 \qquad \text{thus} \\ 7000 &= 13 \text{ cycles} + 1 \text{ Jubilee} + 5 \text{ weeks } (= 6916 + 84) \\ &= 142 \text{ Jubilees} + 6 \text{ weeks} = 13 \text{ cycles} + 12 \text{ weeks} \\ &= 1000 \text{ weeks} = 13 \text{ cycles} + 84 \text{ years}^{11}. \end{aligned}$$

The 84 years which appear here for the completion to 7000 of 13 cycles are called "years of the Evangelists"[12].

Jubilees (from 1 to 10) are occasionally marked on the margins of 532-year tables[13], without becoming a systematically used chronological parameter.

---

[5] Grébaut [1920] p. 218; Upps 3 58^{b}, 10. For the background of this belief in a 7000-year duration of the world, cf. e. g., Weld Blundell, Roy. Chron., p. 496, n. 1. For the role of the year 7000 in the Byzantine church, cf. Piper, Kalend. p. 149: „Das Ende der Ostertafeln und das Ende der Welt." For the association with the 7 days of creation, cf. e. g., Charles-Morfill, Enoch, p. XX.

[6] Dillmann, Br. Mus. Cat. p. 4b (from BM Add 16188). Cf. also below, note 20.

[7] Grébaut [1920] p. 217 f.; Upps 3 62^{a}, 2.

[8] Scaliger, p. 677, 3 f.; Berol 84 22^{a} II, 2 to 4.

[9] Grébaut [1920] p. 219; Dillmann, Br. Mus. Cat. p. 4b; Upps 3 60^{a}, 16 to 60^{b}, 9.

[10] Grébaut [1920] p. 217.

[11] Grébaut [1920] p. 218 f.; Upps 3 57^{b}, 8 to 58^{b}, 2 and CSCO 235, p. 99, 15 f./236 p. 87; Berol 84 22^{a} II, 13 to 22^{b} I, 2; BM Add 24995 27^{a} II, 13—22; Scaliger, p. 676, 18 to 27.

[12] Cf. s. v. Evangelists (p. 128).

[13] Vindob 6 4^{a} to 17^{b}.

The relations between "weeks" and 532-year cycles are expanded into sequences so that n weeks correspond to $m - 1$ cycles plus δ years[14]:

weeks		cycles		δ
$1 \cdot 700$	=	$(2-1) \cdot 532$	+	168
$2 \cdot 700$	=	$(3-1) \cdot 532$	+	336
$3 \cdot 700$	=	$(4-1) \cdot 532$	+	504
$4 \cdot 700$	=	$(6-1) \cdot 532$	+	140
		etc.		
$9 \cdot 700$	=	$(12-1) \cdot 532$	+	448
$10 \cdot 700$	=	7000.		

A similar pattern operates with millennia[15]:

$$1 \cdot 1000 = (2-1) \cdot 532 + 468$$
$$2 \cdot 1000 = (4-1) \cdot 532 + 404$$
etc.
$$6 \cdot 1000 = (12-1) \cdot 532 + 148$$
$$7 \cdot 1000 = 7000.$$

Some texts[16] assemble the 7000 years, following a "Computus of the Sibyl", from 8 "suns" of 778 years each, plus a ninth "sun" 2 years shorter (indeed: $8 \cdot 778 + 776 = 7000$). The term "sun" is here used as an equivalent of zaman "time", "age", a terminology also known from the "Tiburtine Sibyl"[17].

Similar cycles are ascribed to Ezra ($143 \cdot 49 = 7007$) or to Enoch ($10 \cdot 700$)[18]. A "Week of the Prophet Jeremy" of 343 years ($= 7 \cdot 49$) is another modification of the cycle of Jubilees[19]. For the Lord 7000 years are only 7 of his days, whereas for us they correspond to 2556750 days.

---

14 From Berol 84 11ᵃ II, 21 to 11ᵇ I, 21 and CSCO 235, p. 99, 14 to 28 (the translation in CSCO 236 p. 87/88 makes no sense).

15 In the text the millennia are, by mistake, one line out of step.

16 Gróbaut [1920]; Upps 3 60ᵇ, 10—61ᵃ, 7; condensed version in BM Add 16188 (Dillmann, BM Catal. p. 4, slightly garbled) and BM Add 16217 17ᵃ upper margin I/II.

17 Cf. also above, s. v. Feasts, p. 143.

18 E. g. Dillmann, BM Cat. p. 4; Basset, Apocr. X p. 40.

19 E. g. BM Add 19658 62ᵃ II, 2/3; also 58ᵃ 14/15 in connection with a "Week of Daniel".

The years are here taken to be 365¼ days whereas for the cycle of the Sibyl 7000 years are equated with 2548000 days[20]. Hence the years of the Sibyl are the same as the years of Enoch, i. e. 364 days long[21].

20 BM Add 16217 17ᵃ upper margin I/II, 6. Cf. also above note 6.

21 "Weeks" (without numerical detail) are also attested in the Book of Enoch: from 1 to 7 in ch. 92, 1 to 93, 10 and from 8 to 10 in ch. 91, 11 to 17. The Maṣḥafa Milād (15th cent.) assumes unhesitatingly "weeks" of 700 years each. Cf. CSCO 222 p. 56. What the 7th "part" ('eda) of the 10th week represents I do not know. Modern theologians (e. g. Dexinger, Apok.) ignore such technicalities.

Judith

See Feasts (p. 145).

Kēkros and Kentros

"Kēkros" and "kentros", no doubt, are Greek loan-words. It is another matter, however, to understand their meaning as technical terms and to explain their origin. The naive etymology kēkros = circus[1] is not worth discussion beyond remarking that nowhere in ancient astronomy does circus appear as a technical term. On the other hand kentros suggests κέντρον[2], well known in Greek planetary theory as the name for an important parameter. Furthermore the four "κέντρα", i. e. the points of intersection of the ecliptic with the horizon and the meridian, are of great importance in astrological doctrine. Unfortunately neither one of these well defined usages shows any connection with the Ethiopic term. The texts themselves declare both terms as "translations", kēkros meaning "large part", and kentros "small part"[3]. One should not attach too much weight to this kind of scribal explanation.

As we shall see presently, both terms are connected with units of time. These units, however, belong to two widely different backgrounds: the "Enoch" tradition with its 18-division ("parts") of the nychthemeron and the extrema $M : m = 2 : 1$, as distinct from the hellenistic norm of

1 Dillmann, Lex. 859: "sine dubio κίρκος circus i. e. sexagesima diei pars".

2 Dillmann, Lex. 854: "ut videtur quadrantes annorum" (Dillmann seems to know only the plural). This translation is probably suggested by the assumed identity with κέντρα.

3 Vat 119 69ᵃ I, 13; 89ᵇ II, 11/12 and Vat 106 92ᵇ II, 9, respectively.

24 hours and $M : m = 15 : 9$ that characterizes "Greece"[4]. Both traditions overlap in our Ethiopic treatises and make it difficult to come to a clear separation of usages.

Kēkros (*kk*)

The least ambiguous information about the meaning of kēkros comes from diagrams and texts which associate the changing illumination of the waxing or waning moon with a linear sequence which varies between 4 kēkros and 60 kēkros in 15 steps of 4 kēkros each[5]. It is obvious that here the kēkros measures the illuminated fraction of the lunar disk, increasing or decreasing each day by 4/60 of the total area. In terms of the lunar diameter one could say that 1 kēkros represents 1/60 of its length, but this formulation is not explicitly attested in our texts.

The lunar phases can also be described by the time which the moon is visible after sunset, again assumed to vary linearly between zero and 12^h at full moon[6]. Also these daily increments are associated with a sexagesimal division so that

(1a) $4^{kk} \sim 0;48^h, \qquad 60^{kk} \sim 12^h.$

Hence 1 kēkros is here the equivalent of 12 minutes of time in the moon's visibility.

Under these assumptions moonset follows sunset by exactly 4^h at day 5 of the lunar month, exactly 8^h at day 10, and 12^h at full moon at day 15. For the other days fractions of hours are to be added and we have texts[7] which express these additional intervals in terms of kēkros which we can tabulate as follows:

Day			kēkros
1	6	11	1
2	7	12	2
3	8	13	3
4	9	14	4

[4] Cf. s. v. Hours, p. 167.

[5] E. g., BM 809 3^a to 4^a; BM Add 24990 27^b, 28^a; Vat 119 69^a I, 11 to 70^a II, 19; 90^a I, 3 to II, 18. Cf. also s. v. Moon, p. 196.

[6] E. g., Berol 84 29^b I, 20 to III, 4 (incorrectly taking 15 days as 15 hours).

[7] BM 816 17^b I, 1 to 21; BM Add 24995 33^a I, 14—II, 24; Berol 84 30^a I, 7 to 21 (badly garbled).

This agrees with the relation (1 a) which implies that

(1 b) $5^{kk} \sim 1^{h}$.

Another treatise[8] refers this relation to the "computus of the sun and the moon" in contrast to the "computus of the stars" in which

(2 a) $15^{kk} = 1^{h}$

and thus

(2 b) $24^{h} = 360^{kk}$.

Here again we have a parallel to a Greek norm: the division of the day into 360 "time degrees" (χρόνοι)[9]. But also in Ethiopic we have the subdivision of the nychthemeron into 360 "parts" so that kēkros and part would be equivalent units. This would support the interpretation mentioned at the beginning (above, p. 175).

Also in some short Amharic sections (hence late) that were added to our treatises, one finds sexagesimal fractions denoted by kēkros, e. g. 15, 30, 45, and 60 kēkros for the hours 3, 6, 9, and 12 of the day[10]. In connection with Arabic lunar parameters one finds a formulation "3 kēkros, 38 seconds, 3 thirds, 11 fourths" for 0;3,38,3,11 days[11]. Abū Shāker, when discussing the lunar months of 354;22^{d} in the Arabic calendar, calls the days "of first place", the fractions "of second place" or "kēkros"[12].

In all cases discussed so far, kēkros could be rendered as "sixtieths" corresponding to the Greek ἑξήκοντα.

Kentros (*kn*)

The situation is much less clear for the term kentros. In the above mentioned diagram[13] for the lunar phases, one also finds a column "kentros" in which the days of the waning moon are counted from 1 to 15. The translation "day" would seem obvious if we had no other additional usages.

---

[8] Berol 84 33^{a} II, 16 to III, 3 (trivial scribal error: 4^{h} instead of 1^{h}).

[9] Cf. s. v. Hours, p. 168 (1 b).

[10] BM 816 19^{b} II, 12—15. Similarly Marḫa 'E p. 381 II, 20; p. 382 I, 28 et passim.

[11] Vindob 6 3^{a}, 7/8 (for the error of the 19-year cycle). In BN Abb 37 2^{a} I, 19/20 we find the statement "60 kēkros is 1 day". Also the term daqāyq (i. e. Arabic daqīqa, "minutes") is attested as rendering of kēkros: Vindob 6 32^{b} I, 5—12; Vat 54 88^{a}, 9—89^{a}, 19. Parallel: Ambros B 22 182^{a}, 2—182^{b}, 11 and 1^{a}, 1—2^{a}, 2 (incomplete and out of order).

[12] Chapter 11 (BM Add 16252 32^{b}, 22—33^{a}, 5).

[13] BM 816 16^{b}.

A different meaning, however, is required in rules for the variation of the length of daylight. There we find[14] that

$$1^{\text{kn}} = 0;2^{\text{h}}$$

the hours being hellenistic equinoctial hours. It is perhaps only an anachronistic extention if the "hours" (p) of Enoch, 18 of which make one day, are also set into relation to the kentros, such that

$$1^{\text{kn}} = 0;2^{\text{p}}.$$

In this way kentros seems to correspond in both cases to 1/30 of the equinoctial "hour".

Occasionally kēkros and kentros can be combined. We are told, e. g., that the Lord "caused to spread out (?) kentros and their kēkros, making windows where (sun, moon, and stars) are setting and rising in their circuits"[15]. Here one could think that the kēkros are the sexagesimal fractions (minutes) of the kentros.

The rising of the sun in the single gates can be associated with 30 kentros[16], perhaps in the same sense representing 30 days as 15 kentros describe the time of the waxing or waning moon[17]. But the same text seems also to relate 12 kentros with each gate, inexplicable by otherwise known data. Unfortunately these are not the only obscure passages that involve kēkros and kentros[18].

In conclusion one may perhaps say that the meaning of kēkros as 1/60 and of kentros as 1/30 is fairly well established, but neither one of these meanings leads to a reasonable etymology.

---

14 Cf. s. v. Hours, p. 168.

15 BM 693 (Cat. Wright No. 269, p. 179); cf. CSCO 30 p. 113, 14; 31 p. 101.

16 Berol 84 27^{b} I, 1 to II, 10; perhaps also Vat 106 92^{b} II, 6 to 11.

17 Above, p. 177.

18 In BM 816 19^{b} II, 15 we have 4 (what ?) 60 kēkros 240 kentros.

Leap Years

Julian years of the era "A. D." which are divisible by 4 have a 29th of February (otherwise only 28). In the Alexandrian calendar the equivalent of the last day of February is always VII 4. Thus we can also say that for all julian years

(1) March 1 = VII 5.

Conversely: in all Ethiopic eras (J, W, D, and G) the years which are congruent 3 modulo 4 have 6 epagomenal days (otherwise 5). In the julian calendar *ep* 6 is August 29. Thus we can say that the epagomenal day 5 is always August 28.

Example:

J			A.D.			A.D.			J		
1	VII	4	9	Febr.	28	9	Aug.	28	1	ep.	5
2	"	4	10	"	28	10	"	28	2	"	5
3*	"	4	11	"	28	11	"	29	3*	"	6
4	"	4	12*	"	29	12*	"	28	4	"	5
5	"	4	13	"	28	13	"	28	5	"	5

As a result of these relations one sees that Alexandrian dates and julian dates are in a fixed relation to one another, leap year or not, between March 1 and August 28. For the period of Easter, which always belongs to this interval, one therefore has the following rule of transformation:

(2) March 1 = VII (Phamenoth = Magābit) 5
April 1 = VIII (Pharmouthi = Miyāzyā) 6.

Length of Daylight

The ratio $M : m$ of longest to shortest daylight is an important parameter in ancient astronomy since in practice it plays the role which we assign to the geographical latitude φ. Thus $M : m = 7 : 5$ characterizes Alexandria, 3 : 2 Babylon, 15 : 9 Hellespont or Byzantium, etc.[1]. For Ethiopia we have two ratios, 2 : 1 and 15 : 9, both drastically incorrect: the first would correspond to about $\varphi = 48°$ (i. e. the latitude of Vienna), the second to Byzantium. The insistence on one, and often on both of these ratios in all Ethiopic treatises illustrates the total lack of understanding of the most elementary astronomical facts.

---

[1] Cf. e. g., Neugebauer, HAMA, vol. II, p. 706 ff.

The ratio 2 : 1 of longest to shortest daylight is first attested in the Book of Enoch[2]; the ratio 15 : 9 is probably of Greek origin. In both cases a linear variation of the length of daylight during the year is assumed, increasing, respectively decreasing, during six months, always with the constant monthly difference "1". In the case $M : m = 2 : 1$, this leads automatically to a division of the nychthemeron into 18 "parts", in the case of 15 : 9 into 24 "hours"[3].

The abrupt change from increasing to decreasing length of daylight, and vice versa, is again contradictory to all observational experiences but it is nevertheless a commonly used device, familiar everywhere where more advanced mathematical techniques are lacking. For the same reason equinoxes and solstices are simply associated with the same day numbers (25th of IV, VII, X, I)[4], ignoring the epagomenal days.

In the Book of Enoch the months are simply numbered (from 1 to 12) but it follows from the variation of the length of daylight that month 1 lies near the vernal equinox. In treatises which operate with the Ethiopic calendar the equinoctial months are Maskaram (I) and Magābit (VII) as shown in Table I:

Table I

Length of		Months			
day	night	Enoch		Ethiopic	
6	12	9		IV	
7	11	8	10	III	V
8	10	7	11	II	VI
9	9	6	12	I	VII
10	8	5	1	XII	VIII
11	7	4	2	XI	IX
12	6	3		X	

For the scheme which operates with the ratio 15 : 9 only Ethiopic months are attested, again with I and VII at the equinoxes:

---

[2] Ch. 72, 8 to 34; cf. Charles, Enoch, p. 192 to 196; Flemming-Radermacher, Henoch, pp. 92 to 94.

[3] Cf. also s. v. Hours, p. 168.

[4] Cf. s. v. Seasons, p. 204.

Table II

Day	Night	Months	
9	15	IV	
10	14	III	V
11	13	II	VI
12	12	I	VII
13	11	XII	VIII
14	10	XI	IX
15	9	X	

The positions of the equinoxes and solstices in both schemes agree with those shadow tables which assume $U = 9$, $u = 0$ as extreme shadow lengths[5].

In these schemes which are in fact little more than qualitative descriptions, it is of little importance whether one kept the same numbers constant for a whole month or interpolated linearly. In principle, however, linear interpolation seems to have been assumed, at least for the scheme of Table II, because we have statements that the given numbers are valid for the first of the month[6]. Also a special coefficient of interpolation per day in the amount of "1 kentros" is known[7], where 30 "kentros" are 1 hour, the monthly difference in Table II[8]. In one version only the linear changes during 90 days from the equinoxes are described. Also the terminology is of interest in so far as the calendar "day" (ʿelat) is consistently distinguished from the "day" (-light) (maʿālt), in contrast to "night" (lēlit)[9].

Sometimes only the symmetry of the Ethiopian months with respect to the length of daylight and night is described (corresponding to the last columns in Tables I and II) without giving numerical values[10].

---

[5] Cf. s. v. Shadow Tables, p. 212. in contrast to (1) which places the solstices into V and XI, the equinoxes into II and VIII.

[6] Bibl. Nat. Cat. Zotenberg Nos. 126—128 (pp. 152 to 192).

[7] Berol 84 28^{b} II, 14 to 29^{a} I, 1.

[8] Cf. s. v. kēkros and kentros (p. 178).

[9] BM 816 18^{a} I, 15 to II, 17.

[10] Vat 106 92^{b} I, 23 to II, 2.

A good example for the tenacity with which the ratio $M : m = 2 : 1$ for the lengths of daylight was kept alive is a list[11] of "Hebrew" (actually Syriac) and "Roman" months, their lengths in days (as in the julian calendar) and their first and last days in the Ethiopic calendar, followed by the lengths of day and night (reckoned in "parts") and the "gate" of sunrise as if based on Enoch. The scheme for the first three months will suffice as an example:

"Hebrew"	Roman	Days	Ethiopic		"parts"		Gate
			first	last	d.	n.	
Nisan	April	30	$\overline{\text{VIII}}$ 6	$\overline{\text{IX}}$ 5	10	8	4
Ayar	May	31	$\overline{\text{IX}}$ 6	$\overline{\text{X}}$ 6	11	7	5
Ḥaziran	June	30	$\overline{\text{X}}$ 7	$\overline{\text{XI}}$ 6	12	6	6

Textual Evidence

$M : m = 2 : 1$. Berol 84 28^{a} III, 14 to 28^{b} II, 9 (Ethiopic months)
BN 64 37^{a} to 38^{a}; cf. Grébaut [1918, 3]
BM 816 18^{a} II, 17 to 19^{a} I, 4 (daily variation, Ethiopic months)
Vindob 6 1^{a}, 1 to 2^{a}, 21; 3^{a}, 10 to 15 (Syriac and Roman months, gates)

$M : m = 15 : 9$. Berol 84 6^{a} II, 4 to 20; 28^{b} II, 10 to 29^{a} I, 1 (daily variation); 41^{b} I, 1 to II, 20
BN, Cat. Zotenberg Nos. 126 to 128, pp. 152 to 192.
BM 816 18^{a} I, 15 to 25 (daily variation)
Sobhy [1942] cf. s. v. Coptic, Appendix p. 77, Sect. 29
Vat 1 205^{a} I, 17—II, 7

[11] Vindob 6 1^{a}, 1 to 2^{a}, 21 and 3^{a}, 10 to 15.

Luna

See Calendar, Parameters, $\bar{f}$ (p. 46).

Lunar Calendar

See Moon (p. 194).

Lunar Epact

See Maṭqeʿe and Epact (p. 188).

Lunar Mansions

The "lunar mansions" are sections of the ecliptic, obtained by a 28-division, roughly corresponding to the daily progress of the moon. This scheme originated in India and became of importance in Arabic astrology[1]. Owing to Arabic influence the mansions (manāzel) also appear in Ethiopic treatises, their names being close transliterations of the Arabic[2].

The 28-division of the ecliptic leads to 7 mansions for each quadrant and to $2\frac{1}{3}$ mansions for each sign[3]. Another pattern assigns 13 days to each mansion, beginning with Miyāzyā (VIII) 23, with one interval of 14 days (from XII 20 to epag. 4) in order to obtain the total $27 \cdot 13 + 14 = 365$[4]. For this pattern we have a table[5] that associates with the given date the mansion that rises in the morning (A), culminates at midnight (B), and culminates in the morning (C):

	(A)	(B)	(C)
VIII 23	1	15	22
IX 6	2	16	23
IX 19	3	17	24

etc.

where 1, 2, 3, etc. stand for the names of the mansions (beginning with 'esurutayn). This table describes the following situation: when in VIII 23 the sun is near mansion No. 1 (at heliacal rising), then No. 22, one quadrant back, will culminate, while No. 15, which is opposite the sun, will culminate at midnight; 13 days later No. 2 rises heliacally, while No. 23

---

[1] Cf. e. g., Bīrūnī, Astrol. 164 (ed. Wright, p. 81 ff.); Chronol., trsl. Sachau, p. 343 ff.

[2] BM 816 5^{b}—6^{b}; BN 160 74^{a}; Vat 119 62^{b} I, 13—64^{a} II, 18. For modern lists cf., e. g., Ideler, Ursprung und Bedeutung der Sternnamen (1809) pp. 287—289; Kunitzsch, Arabische Sternnamen in Europa (1959) p. 55 f.

[3] In Vat 119 the mansions are also associated with the 28 letters of the Arabic alphabet (in the order of the alphabetic numerals).

[4] Since the mansions begin with Aries, VIII 23 seems to be the date of the vernal equinox, slightly different from the ordinary norm VIII 25 (cf. s. v. Seasons, p. 204).

[5] BM 816 $5^{b}/6^{a}$.

culminates, and No. 16 is diametrically opposite to No. 2 and the sun. All this is, of course, only a simplifying schematization of the stellar phases during the rotation of the ecliptic.

Schematic pictures of the constellations that constitute the mansions are shown in the second ring of the diagram in BN 160 fol. 74[a]. The stars are represented by dots, numbering between 1 and 14[6].

Obviously the Ethiopic scribes did not always understand what the 28 mansions actually represented. Hence they associated the 28-division with the 364-day year of Enoch by saying that 28 stars rule for 13 days each[7].

Modern scholars are not always much better. Charles, e. g., translates "And the dominion of Elgûfr from Manâzel was then . . . the 1947th year of Alexander . . ."[8] instead of "in the ascendant of the mansions was Elgufr" (without any connection to the era Alexander)[9].

[6] Similar representation for the Decans are found on Egyptian monuments; cf. Neugebauer-Parker, Egyptian Astronomical Texts, Providence 1969, vol. III, 2, Pl. 3, 18, 20.

[7] Vindob 6 32[b] I, 16—18.

[8] Charles, Chronicle, p. 202. This after Zotenberg [1883] p. 587 had given the basically correct rendering "sous la mansion Alghafr".

[9] From the colophon of the translator of the Chronicle of John, Bishop of Nikiu. Indeed, the date II 22 belongs to the heliacal rising of No. 15 at II 21 in the table of BM 816.

Lunar Months

See Moon (p. 194).

Lunar Year

See Moon (p. 194).

Magic

Magical practices are widespread in Ethiopia. In the present context, however, we need not describe details beyond stressing one point: magic has nothing to do with astrology. The purpose of magic is to influence events, to exercise power, or to protect. The goal of astrology is the prediction of the inevitable future on the basis of a given astronomical situation. As in all human affairs the limits are not absolutely rigid, e. g., when "Katarchic" astrology allows for preventive measures; but the basic doctrines as well as the techniques of the practitioners of magic and of astrology remain fundamentally different.

This distinction is also of importance historically. Astrology appears in Ethiopia only late and in Arabic versions. Magic, on the contrary, is deeply associated with Judeo-Christian rites, with the belief in demons and angels, and in the power of saints and rituals. Many of the magical procedures are inherited from pagan, hellenistic, or oriental practices, as is evident, e. g., from a comparison of Ethiopic magical designs with the abundant material available in Greek papyri[1].

There exists a huge modern literature on Ethiopic magic, now easily accessible in the bibliographies of Strelcyn [1955] pp. 483—498, or Lifchitz [1940] pp. 245—250.

Even calendaric concepts found applications in divination and numerology. In all cases known to me, reductions modulo 7 (or multiples of 7) play a role that indicates a connection with the days of the week. The introduction of ṭentyon and epact into these practices points in the same direction.

A simple example is the determination of rainstorms[2]: compute the residue of

$$\text{(1)} \qquad \text{ṭentyon} + 5 \text{ epag. days} + 1 \text{ extra day} \equiv r \mod 7.$$

If $r = 1$, there will be ten downpours; if $r = 2$, twenty, etc., up to seventy at $r = 7$.

More difficult to explain is the meaning of a quantity called barakat, i. e., "blessing" or "prosperity" (β). We have different rules for the computation of β, all of which involve the epact (e). In one text[3] two equivalent definitions are given

$$\text{(2)} \qquad \beta \equiv e + 4 \quad \text{and} \quad \beta \equiv e + 11 \mod 7.$$

The second version could imply that the epact of the coming year is in-

---

[1] Cf. Preisendanz, Papyri Graecae Magicae, 2nd ed., in 3 vols. (1973 ff.); cf. also Vat. Cat. p. 515 f., Nos. 38, 40 to 44, etc. An amusing example of the attitude of a great scholar of the 19th century is the following statement: "I negri, come è noto, sono profondamente superstiziosi. Gli Abissini . . . non facevano eccezione alla regola" (Conti Rossini [1899, 1] p. 203).

[2] Grébaut in ROC 18 (1913) p. 99 f., rendering ṭentyon as "le (jour) initial des (pluies)"! A parallel version is found in Berol 994 66ᵃ II, 4—14, speaking about a "Computus of keramet", i. e. of the rainy season.

[3] Vat 43 250ᵇ I, 10—II, 9. The first formula in (2) is also given in Vat 18

volved. In two other texts[4] three quantities are added, reminiscent of (1):

(3) $\beta = e + 6$ manfaqa $+ 5$ epag. days mod 7

which might be an equivalent to the $e + 11$ in (2). The term manfaqa, "half", could indicate $6 = \frac{1}{2} \cdot 12$ months[5] and thus relate (3) again to the end of the year. For applications of this number, cf. above s. v. Gematria, p. 162.

(Vat. Catal. p. 79, 27).

[4] Berol 84 24[a] II, 8—13 and Vat 69 134[b], 4—8.

[5] Cf. s. v. Feasts, p. 139.

Maps, Rose of Winds

Circular maps and diagrams are not rare in Greek treatises, e. g. in the Handy Tables[1]. It is therefore not surprising also to find similar diagrams in Ethiopic texts. The following examples may suffice:

Frankfurt 18, Goldschmidt, Ab. Hs. p. 67 and p. 103. Fol. 200[b] a "Wheel of Winds" (śaragalā nafās); cf. below, s. v. Orientation, Fig. 4, p. 199. Fol. 201[a] shows a circular map with Axum at the center of two concentric rings. The outer ring, divided into 14 compartments of uneven size (two left empty), gives the names of provinces in approximate relation to Axum; the inner ring is a rose of winds as on the preceding folio.

Ṭānā 34 fol. 19[b] (Hammerschmidt I, p. 158 V). Parallel to the map Frankfurt 18, fol. 201[a].

Jerusalem 268, fol. 3 (Littmann [1902]); parallels: BM Add 16217 24[a]; EMML 49, 28[a]; EMML 215, 32[a]. A rectangular arrangement, probably derived from a circular diagram. Center ḍa, i. e. "sun". The surrounding 8 fields give directions as in a rose of winds. The next outer ring of 12 fields lists the winds and their qualities. Four additional fields in the corners name the four rivers of Paradise and the Evangelists.

An ad hoc invention, with rather implausible details, are the grotesque maps given by Milik, Enoch, p. 18 and p. 40.

[1] Cf., e. g., Neugebauer, HAMA III, p. 1139 s. v. circular diagrams; also Neugebauer [1975].

Maṣḥafa

See Books (p. 26).

Maskaram

Meaning: rainy season (cf. Dillmann, Lex. 834 f.). First month of the Ethiopic year, identical with Alexandrian Thoth. Hence Maskaram 1 = Thoth 1 = August 29 (or 30, for the years with $\varepsilon \equiv 0$) in the julian calendar.

Mathematics

Not a single mathematical treatise, or reference to mathematics in general, has been found (at least to my knowledge) in the huge amount of Ethiopic texts described in our catalogues.

Maṭqeʿe and Epact

Maṭqeʿe and epact are the foundation of the whole computus
BM Add 24995 21 [b] I, 15/16

The maṭqeʿe (m) gives the date in the Ethiopic civil calendar of the Jewish New Year's Day[1]. According to Jewish tradition this date represents a (schematic) new moon's date that falls near the autumnal equinox. It therefore belongs to one of the first two months of the Ethiopic year and it always falls later than the Ethiopic Maskaram 1.

In order to keep the beginning of the Jewish lunar year near the autumnal equinox, the dates m are bounded by fixed dates within the Ethiopic calendar (which, in principle, is based on the solar year):

(1) $$\text{I } 15 \leqq m \leqq \text{II } 13.$$

This interval contains the autumnal equinox at I 25[2]. Since the interval (1) is shorter than 30 days, the day number of m determines uniquely the month of New Year's Day[3].

---

[1] For the etymology of maṭqeʿe, cf. Dillmann, Lex. 1224 f. (with a meaning probably similar to the Jewish shofar, a horn or trumpet-like instrument).

[2] Cf. s. v. Seasons, p. 204.

[3] In Upps 3 73 [a], 8 to 10 it is stated that m cannot fall on I 30 or II 1. This in fact is incorrect for I 30 but correct for II 1. Since m is periodic with 19 years, it is clear that $30 - 19 = 11$ dates must be omitted; cf. s. v. Calendar, p. 59, Table "XIX".

A "lunar year" of 12 lunar months, i. e. of 354 days, is 11 days shorter than the "solar year" of 365 days. Assume that in a certain year m has the extremal position II 13. The next year the corresponding new moon will fall 11 days earlier in the solar year, thus m = II 2. In the third year the date will be I 21, but in the fourth year the new moon at I 10 can no longer start the lunar year because of the lower boundary prescribed in (1). Hence we add one whole month and make II 10 the first new moon of the year. Thereafter I 29 and I 18 are permitted, but I 7 is to be replaced by II 7.

In this way the sequence of the civil dates m of the lunar New Years days can be established by subtracting 11 (or adding $30-11 = 19$) step by step within the permitted limits (1). Instead of operating in this fashion with m one can also use the numbers

(2) $$e = 30 - m$$

which are called "epact" ('abaqtē)[4]. Note that m represents calendar dates while the e are simple numbers[5]. In our example we have[6]

(3)

m = II 13	$e = 17$	
II 2	28	
I 21	9	
II 10	20	etc.

Hence e either increases by 11 or decreases by 19.

Babylonian astronomers made use of the fact that $12 \cdot 19 + 7 = 235$ mean synodic months are very nearly the equivalent of 19 sidereal years. Undoubtedly on the basis of this knowledge, widespread in antiquity, it was also obvious that the numbers m and e had to be repeated with a period of 19 years if they should provide the dates for a sequence of new moons. Continuing procedure (3), however, we would end up after 18 steps with $e = 5$ and the next step would bring us to $e = 16$ instead

[4] Also called "lunar epact" ('abaqtē warḫ), in contrast to "solar epact" which is the ṭentyon. Ṭānā 34 100ª I, 8 explains 'abaqtē as meaning "remainder of the year". Chaine's statement (Chron. p. 105, n. 1), "On trouve fréquemment . . . la formule $e = 29$, $m = 30$" is simply wrong.

[5] Since $e+1$ is the age of the moon in the last day of the month that contains the maṭqe'e and since VII 26 = I 206, one finds for VII 26 as the age of the moon $206 - 30 + e + 1 = 177 + e$. But $177 = 3 \cdot 59$ and therefore the age of the moon at VII 26 (= March 22) is e. It is this medieval definition of the "epact" which is used, e. g., by van de Vyver [1957]. I have never seen it in the Ethiopic (i. e. Alexandrian) computus.

[6] Similarly in Vat 18 251ᵇ, 1—252ᵇ, 2 (Vat. Catal. p. 80, 31).

of back to the initial $e = 17$. Obviously this is caused by the fact that $11 \cdot 19 = 209 \equiv -1 \mod 30$; therefore one had to introduce into every 19-year cycle a correction in the form that e increases once by 12 instead of by 11 as in all other steps. In the late Middle Ages the European computists called this correction the "saltus lunae". Astronomically it reflects the fact that the 11-day increment is based on a mean synodic month of exactly 29½ days in length whereas the 235 months in 19 years require a slightly longer mean synodic month. But no accurate astronomical knowledge is required to assure the periodicity of the epacts in 19 steps: the simple device of the saltus lunae produces the desired result. Hence we find in the Ethiopic calendar the following sequence for the epact e, repeated in each 19-year cycle:

$$e = 0 \quad 11 \quad 22 \quad 3 \quad \ldots \quad 7 \quad 18 \tag{4a}$$

in which the saltus lunae is placed between the last and the first year of consecutive cycles[7].

The use of the sequence (4a) is well attested also in Byzantine astronomical tables. Since these tables are frequently based on the Era Diocletian a comparison with the Ethiopic pattern is a simple matter. We know, e. g., that the "years of Grace" (the era G) begin their first cycle with the year Diocletian 77[8]. The Byzantine tables[9] give for the "epact" (ἐπακτή) the sequence

D:	76	77	78	79	
e:	18	29	11	22	etc.

whereas the Ethiopic tables give

G:	0	1	2	3	
e:	18	0	11	22	etc.

The only difference consists in the location of the saltus lunae one year later in D than in G. The norm in G is probably motivated by the convenience of beginning a cycle with $e = 0$. This leads for m to the sequence

$$m = 30 \quad 19 \quad 8 \quad 27 \quad \ldots \quad 23 \quad 12 \tag{4b}$$

in each 19-year cycle[10].

---

[7] It is, of course, irrelevant where one locates the "saltus lunae" within a cycle. The European computists (and some modern historians) act as if this choice were a matter of great significance.

[8] Cf. s. v. Era (p. 117, Table I).

[9] E. g., Vat. gr. 1291 fol. 47[a] ("Handy Tables").

[10] Cf. s. v. Calendar, App. I, p. 59, Table "XIX".

The direct computation of e or m for any given year is a simple matter. If c is the cycle number in the 19-year cycle for the year in question, it follows from (4a) and (4b) that

$$(5\text{a}) \qquad e \equiv (c-1) \cdot 11 \qquad m \equiv (c-1) \cdot 19 \mod 30.$$

Since corresponding year numbers in one of the eras W, G, or D are congruent modulo 19, one obtains for a year E of any one of these eras[11] the cycle number c from

$$(5\text{b}) \qquad c \equiv \text{E} \mod 19.$$

Hence e and m can be found[12], satisfying (2).

The importance of the parameters e and m lies in the fact that they determine all dates in the Jewish lunar calendar[13]. For example, the date of Passover (p) is related to m by the rule

$$(6\text{a}) \qquad p = m + 190 \equiv m + 10 \mod 30$$

hence with (2)

$$(6\text{b}) \qquad p + e \equiv 10 \mod 30$$

which allows one to find immediately the date of Passover for a year of given epact e.

Since to the civil date m corresponds a lunar date 1, it follows from the basic relation (2) that the first day of the next civil month has the lunar date $\bar{\nu} = e + 2$. Indeed, this parameter is tabulated in the "re-arranged" 19-year tables[14] and explicitly characterized as a lunar date (śarqa lēlit).

Strictly speaking $e + 2$ gives the lunar date of II 1 or III 1 depending on m belonging to I or II. Assuming, however, the usual schematic "months" of 30 days, $e + 2$ will also be the lunar date of I 1[15]. This explains, e. g., entries in the annals of 'Īāsu II[16]. There we find listed for the ten years W 7223 to 7232 (A. D. 1730 to 1740) the parameters ε, e,

---

11 This does not hold for the Era of Incarnation.

12 These rules (5) are found in many treatises, e. g. Scal p. 673 1 to 16; Upps 3 65^{b}, 4—13; 66^{a}, 5—13; 65^{b}, 13—66^{a}, 5; 66^{a}, 13—66^{b}, 5; Vat 119 85^{a} I, 9 to II, 6 (credited to the Patriarch Demetrius [A. D. 214/5]); 86^{a} I, 15 to II, 8.

13 Cf. s. v. Feasts, p. 131.

14 Cf. s. v. Calendar, Tables p. 36.

15 The maximum error caused by this assumption is one day within the schematic lunar calendar.

16 CSCO 61/66, p. 26/27 to p. 87/94.

and m[17] followed by the remark "on $e+2$ in evening epoch, on 1 in morning, epoch, . . ., the beginning of the month Maskaram".

In the 532-year tables one finds a column (*le* usually the third one) which is connected with the epact by its heading "beginning of epact" (ledata 'abaqtē). A connection with e, or rather with m, is obvious since the numbers *le*, similar to the numbers m, increase from year to year by 19 (modulo 30) as the comparison of the two sequences shows (for $c = 1, 2, \ldots$)

$$(7) \qquad \begin{array}{lrrrrl} m = & 30 & 19 & 8 & 27 & \ldots \\ le = & 0 & 25 & 14 & 3 & \ldots \end{array}$$

Thus, excepting the inconsistency in the first step of *le*, one has simply

$$(8) \qquad le \equiv m + 6 \mod 30.$$

The inconsistency in the first step can be eliminated if we associate both sequences with their calendar dates[18] and interpret 0 as I 1 (or epagomenal day 6). Then we obtain

$$(9) \qquad \begin{array}{rlllll} m = & \text{I } 30 & \text{I } 19 & \text{II } 8 & \text{I } 27 & \ldots \\ le = & \text{I } 1 & \text{XII } 25 & \text{XII } 14 & \text{ep. } 3 & \ldots \quad \text{etc.} \\ m - le & 29^{d} & 29 & 59 & 29 & \end{array}$$

This shows that the dates *le*, if restricted to the interval

$$(10) \qquad \text{XII } 8 \leqq le \leqq \text{I } 1 \qquad \text{(or ep. 6)}$$

as m is restricted to (1):

$$\text{I } 15 \leqq m \leqq \text{II } 13,$$

precede the Jewish New Years Day either by one or by two lunar months. Thus *le* gives the civil date of the last new moon that falls into the preceding year[19]. This interpretation is supported by tables which relate alternating full and hollow months to the civil calendar[20].

---

[17] At the year 7225 (p. 49/52) read 16 for $e + 2$, not 17 (edition). At the year 7226 (p. 81/86) the numbers of e and m are interchanged and $m + 2$ is taken instead of $e + 2$. All this is overlooked in the edition.

[18] These dates are listed (with occasional disturbances at the beginning or at the end) in Berol 84 2ᵃ I, 15—23; BM Add 24995 22ᵃ II, 21—22ᵇ I, 21; EMML 2077 103ᵃ II, 7—32; Princ 5884 6ᵇ I, 10—II, 16; Ṭānā 34 100ᵃ I, 23—II, 14, Ṭānā 53 107ᵃ I, 14 II, 11; Vat 119 85ᵇ I, 16—II, 11.

[19] Weld Blundell (Roy. Chron., p. 505) probably had a relationship of this kind in mind when he said "the Birth of Epact (as he translates ledata 'abaqtē) falling in one year corresponds to the Epact of the next year".

[20] EMML 49 85ᵃ—86ᵃ and EMML 215 98ᵃ—99ᵃ (close parallels).

Much confusion has been caused in calendaric treatises by mixing up the 11-day epact of the 19-year cycle with the "epact" of 10 days which would result from the 364-day year of Enoch[21], further complicated by the Alexandrian leap year pattern with either 5 or 6 epagomenal days. The name of Enoch is often invoked in these discussions, obviously without actual historical basis[22].

21 Book of Jubilees, VI, 36; also Berol 84 35ᵃ I, 13 to 19.

22 "Sayings of Enoch" in Berol 84 41ᵃ I, 26 to 41ᵇ III, 14 (parallel for the later part in Princ 5884 14ᵃ I, 1 to 14ᵇ III, 10); reference to the Book of Jubilees: 41ᵃ II, 30/31; also in EMML 215 66ᵇ I, 28.

Months

We find three types of months in our material: (a) the months of the Alexandrian-Coptic-Ethiopic calendar, (b) schematic lunar months of alternating 29 and 30 days' length, (c) the months of the Enoch calendar. The months of the Alexandrian calendar are always 30 days long, excepting 5 epagomenal days at the end of three consecutive years, 6 in every fourth year. These years begin with

(1) Month I = Alexandrian Thoth = Maskaram.

Maskaram 1 coincides three times with August 29, once with August 30.

It is convenient for calendaric computations to count days from day 0 of Maskaram based on the following table:

	Day 0		Day 0		Day 0
Maskaram (I)	0	Ṭer (V)	120	Genbot (IX)	240
Ṭeqemt (II)	30	Yakātit (VI)	150	Sanē (X)	270
Ḫedar (III)	60	Magābit (VII)	180	Ḥamlē (XI)	300
Tāḫšāś (IV)	90	Miyāzyā (VIII)	210	Naḥasē (XII)	330
				paguemēn (ep.)	360

The origin and meaning of the Ethiopic names of the months seems to be unknown, except for Maskaram ~ rain or winter, and Ḥamlē ~ harvest[1]; Ḫedar probably comes from Coptic/Greek Hatūr/Ἀθύρ.

1 Dillmann, Lex. 834 and 70 f. Also Guidi, Voc. (562 s. v. war) does not give etymologies or explanations.

The schematic lunar months, alternating between hollow and full, underlie the Jewish calendar which is embedded, e. g., in the 532-year tables (characterized by the term "evening epoch")[2]. The same pattern also provides the framework for the time intervals assumed for the risings and settings of sun and moon through the "gates"[3] of the horizon and thus belongs to the astronomical concepts typical for the Book of Enoch[4].

It is not surprising to find references to schematic lunar months in other contexts as well[5]. For example, at the beginning of the World, months begin and end in relation to the civil calendar as follows[6]:

month I	begins: I, 1	ends: I, 29	thus hollow
II	I, 30	II, 29	full
III	II, 30	III, 28	hollow

etc. Here, as frequently in similar cases, the same names are used for the civil months and for the schematic lunar months which only partially coincide with them.

In the Book of Enoch the months are arranged in four equal seasons of 3 months each, of $30+30+31$ days in length[7]. No names are given to the months but it follows from the context that the first month belongs to the Spring season[8]. This is confirmed by other treatises which count "month 1 (i. e.) Miyāzyā (VIII), month 2 (i. e.) Genbot (IX)" etc.[9]

It is, however, important to note that our treatises frequently identify the counting of Enoch's months with the Ethiopian months, i. e. equate month 1 with Maskaram, 2 with Ṭeqemt, etc. This terminology is bound to produce misunderstandings[10].

---

[2] Cf. s. v. Calendar (p. 32); Feasts (p. 139).

[3] Cf. s. v. Gates (p 159 and Table I, p. 160).

[4] This suggested to Charles (Enoch, p. 190) the following amusing remark: "He [the author of Enoch] knows apparently the length of the synodic months . . . which was not published till the time of Gamaliel II, 80 to 115 A. D.".

[5] E. g. Vat 106 92[b] II, 2 to 4; Berol 84 29[b] I, 3 to 5; 11/12; 31[b] III, 5 to 11.

[6] Berol 84 34[b] II, 4 to 39[a] III, 26.

[7] Enoch, ch. 82, 11 to 15. Also Vat 106 91[b] I, 7 to 92[b] I, 23; duplicate Vat 119 66[a] I, 1 to 69[a] I, 10.

[8] Enoch, ch. 72, 6/7.

[9] Berol 84 40[b] III, 27 to 41[a] I, 26.

[10] Cf. s. v. Seasons, p. 206.

There is no textual evidence whatsoever for the invention by modern scholars of a "month" of 28 days and a "liturgical year" of always 13 such months in order to explain the 364-day year of Enoch[11]. It is equally incorrect (and a modern invention) to call the epagomenal days a "13th month".

References to various calendars are frequent in Ethiopic treatises but often incorrect. The Syriac months, e. g., are called "Hebrew" months[12]; Syriac and Roman months are identified, or "Greek, Coptic, and Arabic" names are actually only variants of Coptic[13]. Also the equivalences between different calendars are often in disorder and in many cases make comparative lists worthless.

Syriac months appear also in diagrams which represent lunar phases, gates, orientations and winds, etc., in combination with zodiacal signs, as usual in Arabic terminology[14].

---

[11] Cf. e. g., Charles, Jubilees, Introd., p. LXVIII; p. 55, notes 29—30.

[12] Vindob 6 1[a]/3[a].

[13] Princ 5884 12[a] II, 12 to 12[b] II, 3.

[14] Cf. s. v. Diagrams, p. 95.

Moon

In Babylonian astronomy, lunar theory occupies a central position, its main goal being the accurate prediction of the moon's first and last visibility on which the true lunar calendar depends. This problem vanishes from Greek astronomy, which operates exclusively with the schematic months of the Egyptian calendar. Ptolemy, for example, in the Almagest completely ignores the problem of lunar visibility (in marked contrast to planetary visibilities).

Nevertheless, the 19-year cycle, usually associated with the name of Meton (of the 5th century B. C.), remains well known. Diodorus (1st century B. C.), Geminus (1st century A. D.), and Censorinus (3rd century A. D.) all praise its superior quality[1]. For Greek lunar calendars the 19-year cycle never became of practical importance. In the cuneiform texts, however, from the Persian period on, the calendar was strictly governed by this cycle and it seems a plausible assumption that the Jewish adherence to the same lunar calendar is due to Mesopotamian in-

---

[1] Diodorus XII 36, 2; Geminus, Isag., VIII, 48; Censorinus, De die nat., XVIII, 8. It also should be remembered that astronomical cuneiform texts were still computed as late as A. D. 74/75 (cf. Sachs [1976]).

fluence. At any rate we know, e. g., from Anatolius (3rd century A. D.) that the Jews relied on the 19-year cycle from pre-Christian times onward[2]. And, as is well known, the early Easter computus followed the Jewish procedure in determining the 14th of Nisan.

One must not assume, however, that the Jews applied anything like the sophisticated Babylonian lunar theory to the determination of the syzygies. It is clear that the simple 19-year calendaric pattern was considered adequate to establish the dates of the lunar phases within the civil (e. g. Alexandrian) calendar. In this framework a simple pattern of alternating full and hollow months sufficed. The following sequence of lunar phases may serve as an example[3]:

	invisible:	I 8	II 8	III 7	IV 7	
(1a)	first crescent:	I 9	II 9	III 8	IV 8	etc.
	full moon:	I 23	II 23	III 22	IV 22	

The dates (in the civil calendar) show that corresponding phases are separated by 30, 29, 30, . . . days. The first crescent is given a date one day later than invisibility and full moon follows invisibility after 15 days. The dates in (1a) are said in the text to be upper bounds. A similar scheme is given for the lower bounds:

	invisibility:	I 10	II 10	III 9	
(1b)	first crescent:	I 11	II 11	III 10	etc.
	full moon:	I 25	II 25	III 24	

This indicates that not only an interval of 30 days is possible, e. g. between two consecutive first crescents (as I 9 and II 9) but that it may be as low as 28 days (from I 11 to II 9). This is a faint reflection of the actual variability of the sequence of full and hollow months, well known in Babylonian astronomy[4].

A linear scheme for the increase of the illuminated part of the waxing moon and of the obscured part of the waning moon, both progressing with 1/14 of the area of the lunar disc every day, is found in the Aramaic version of the astronomical Book of Enoch[5]. The waxing moon shines with 1/14 of its total light on the evening of day 1 of the lunar month and reaches totality (14/14) on day 14. The waning moon has lost 1/14 of its light on the morning of day 16 and is totally invisible (14/14 ob-

---

[2] Eusebius HE VII 32, 14—16 (ed. Bardy II, p. 225 f.).

[3] Berol 84 39[b] III, 6 to 40[b] I, 16.

[4] Cf. e. g., Neugebauer, ACT III Pl. 140 to 150.

[5] Milik, Enoch, p. 280/281. For a similar Hebrew version, cf. Milik, p. 68/69.

scuration) on day 29[6]. The Ethiopic version of this description is badly garbled[7] but describes the same sequences.

The two linear sequences of 14 days increasing and 14 days decreasing illumination vary between the extrema 0 and 1, which remain valid for one day each[8]—a pattern well known from linear schemes in Babylonian, Greek, and medieval astronomy.

In Ethiopic texts one also finds a description of the variable lunar illumination in terms of a sexagesimal linear sequence, presumably the result of hellenistic influences. The increment is assumed to be "4 kēkros" (i. e. 0;4)[9] or "1 part" (kefel or 'ede, i. e. 1/15)[10] per day. Consequently the increase and the decrease is now made to last 15 days each.

Already in the Book of Enoch sun and moon are said to be of equal (apparent) size[11]. The same source postulates that the brightness of the sun is seven times the brightness of the full moon[12]. This assumption gives rise to another linear sequence which compares the brightness of the lunar phases with the brightness of the sun:

	day	kēkros	sun
	1	4	105
	2	8	98
(2)	3	12	91
		etc.	
	15	60	7

[6] Milik's translation is almost unintelligible (perhaps in part because of the translation from French into English). The general pattern of the text can be paraphrased as follows: Day 9: the moon shines with 4½ sevenths (= 9/14) of its light, the obscuration decreases 2½ sevenths (5/14) during the rest of the night; during this day (the illumination) increase to 5/7 (= 10/14), keeping (in darkness) a remainder of 2/7 (= 4/14). Similarly for the waning moon: Day 26: obscured, i. e. subtracted from the moon's light, 5½ sevenths (= 11/14), it shines during the rest of the night with 1½ sevenths (= 3/14); during this day (the obscuration) increases to 6/7 (= 12/14), 1/7 (= 2/14) being invisible during the rest of the day.

[7] Enoch, ch. 73, 5 to 8.

[8] Strictly speaking the extrema cover 1½ days (e. g. between last visibility in the morning and first visibility on the next evening) while the linear parts cover 13½ days each.

[9] Berol 84 29ᵇ I, 20 to 30ᵃ II, 18 (with many scribal errors); similarly in the diagrams of BM 809 3ᵃ to 4ᵃ; BM 816 16ᵃ/16ᵇ, 17ᵇ. Cf. s. v. "kēkros and kentros" (p. 176).

[10] Berol 84 30ᵃ I, 22 to II, 18.

[11] Enoch, ch. 72, 37; 78, 3.

[12] Enoch, ch. 72, 37; 73, 3; 78, 4. Slavonic Enoch XI, 2 (Vaillant, p. 11/13). Similarly, Berol 84 29ᵃ II, 12; III, 1 to 3.

Here the last column tells us that the brightness of the sun is $105 = 15 \cdot 7$ times the brightness of the moon on the first day of the lunar month; $98 = 14 \cdot 7$ times on the second day, to 7 times at full moon[13].

It is clear that the arithmetical patterns which describe the variations of the lunar phases, sexagesimally or not, assume a schematic month of 30 days. In another context, however,—e. g. for the lunar calendar of Passover[14] or for the conjunctions of sun and moon in the "gates"—a lunar year of alternating full and hollow months is assumed[15]. Such flagrant inconsistencies did not disturb the men who wrote or used these treatises.

One of our texts[16] even investigates the schematic lengths of the lunar years which one has to assume in order to produce a 19-year cycle (cf. Table I). Besides lunar years of 354 days (i. e. alternation between hollow and full months in that order) one needs lunar years of 355 days by making the first month full, and finally intercalary years of 383 days with a hollow 13th month. The total then amounts to $19 \cdot 365 = 6935$ days. Here, in order to obtain periodicity in 19 steps the "years" are Egyptian years in contrast to the scheme discussed s. v. Cycles (p. 80 Table I), which operates with Alexandrian years and with a fixed difference of 11 for the epacts (plus 12 for the saltus lunae).

Table I

Year	Beginning	Δ	Year	Beginning	Δ
1	I 1		11	I 10	355*
2	I 19	*383	12	I 28	*383
3	I 8	354	13	I 17	354
4	I 26	*383	14	I 7	355*
5	I 15	354	15	I 25	*383
6	I 5	355*	16	I 14	354
7	I 23	*383	17	I 4	355*
8	I 12	354	18	I 22	*383
9	I 2	355*	19	I 12	355*
10	I 20	*383	1	I 1	354

13 DM Add 24090 27^b; BM 809 3^a. Similarly, Vat 119 69^a II, 15 to 70^a II, 19; parallel version 90^a I, 3 to II, 15.

14 Cf. s. v. Feasts, p. 140. 15 Cf. s. v. Gates, p. 160, Table I.

16 Berol 84 34^b II, 4 to 39^a III, 28. Several scribal errors are corrected in our table.

Morning Epoch

See Feasts, Lunar Dates (p. 139).

Nineveh

See Calendar, Tables (p. 49).

Nova

In the historical column ("tārik") of some 532-year tables[1] one finds at $k = 195$ the note "a star became visible". Since these tables belong to the cycle G_3 the year is J 1611, i. e., A. D. 1618/19. This date agrees with Chinese records which report a nova that appeared in 1618[2]. Unfortunately neither one of these sources gives any details.

[1] Berol 994 47[b]; BM 815 8[a] ~ BM 827 109[a].
[2] Lundmark [1921], p. 234, No. 52.

Numerology

See Magic (p. 185).

Numerological Cycles

See Jubilees (p. 173).

Octaeteris

See Cycles (p. 83).

Orientation

The Ethiopic terms for directions are not always unambiguously determined by the context. In particular, the pair north/south is said to have, in the course of time, reversed its meaning[1] and it is therefore advisable to present the evidence on which the interpretation of directional terms must rest.

Of basic importance is the fact that all terms appear in pairs. Among these pairs meśerāq and meʿerāb are unquestionably East and West since their literal meaning is "oriens" and "occidens", respectively[2]. Also

[1] Dillmann, Lex. 334 (samēn) and 1106 (dabub); Abbadie, Cat. p. 156.
[2] For East one has also "morning" (ṣebāḥ).

secure is the literal meaning "right/left" for yemen/ṣegem: in relation to the "gates" on the Eastern horizon "right" must represent a southerly, "left" a northerly direction[3].

In our calendaric-chronological texts we have a number of diagrams, or even schematic maps, that establish beyond any doubt the terminology used in each case for the different directions. All these sources agree, without exception, in the terminology for the cardinal directions: meśeraq/meʿerāb for East/West and orthogonal to it samēn/d̆abub for North/South. The latter pair is also used in the same sense in connection with the "gates". Hence I see no evidence for any change in the terminology for the cardinal directions.

The opposite situation prevails, however, for the intermediate directions. For these we have two pairs of terms: one bāḥr (B) / libā (Λ), the other maseʿe (*ms*) / azēb (*az*). Fig. 4 illustrates the different ways these directions are used in relation to the fixed cardinal directions[4]. It

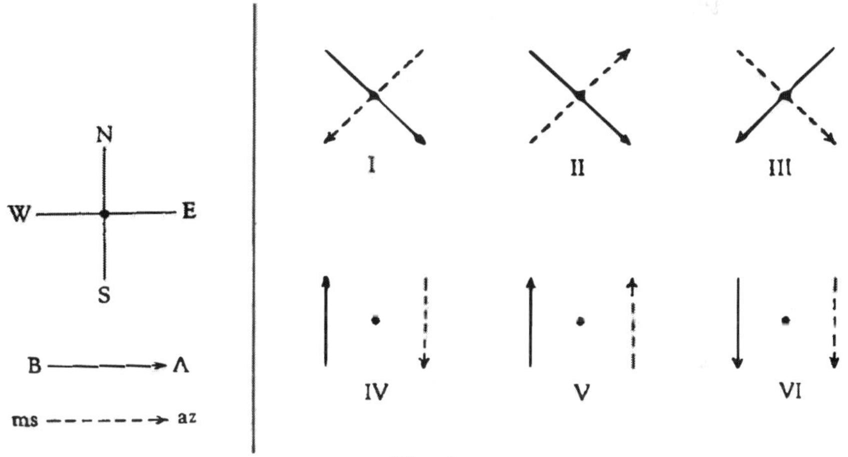

Fig. 4

[3] Cf. e. g., Berol 84 31[a] III, 9—31[b] I, 18.

[4] Textual evidence:

I. Frankf 18 200[b]; BM 812 2[a] (Wright, Cat. No. 384, p. 291); also Guidi, Voc. Am. 94.

II. Frankf 18 201[a]; Ṭānā 34 19[b].

III. EMML 49 27[b] ~ EMML 215 31[b].

IV. Jerus 268, 3; EMML 49 28[b] ~ EMML 215 32[b]; EMML 49 28[a] ~ EMML 215 32[a] (incomplete).

V. Berol 994 68[b].

VI. BM 809 3[a]—4[a].

Note that the same manuscript may contain contradictory terminologies.

is evident from these data that our texts no longer have any clear idea about the four terms referring to the directions between the cardinal points.

Also geographical considerations make no sense: bāḥr, "sea, ocean", (endpoint B of the solid arrows in Fig. 4) can be located in three of the four quadrants. Its opposite element libā, presumably from the Greek wind-direction λίψ (southwest)[5] shows no consistency. But even for the cardinal directions can one find an ambiguous terminology. In biblical usage the pair maseʿe/ʾazēb stands for the direction north/south[6], a terminology attested also in connection with never setting stars "in northerly (maseʿe) direction"[7]. Also the cardinal directions for the winds can combine ṣebāḥ/me ʿerāb with maseʿe/ʾazēb[8]. Evidence for a shifting terminology within the intermediate direction is also found in a gloss in the Book of Enoch (ch. 76, 10) which says "the wind from maseʿe had also the name bāḥr[9]". In Fig. 4 I and V or IV and III explains such a statement. All available evidence seems to demonstrate stability for the cardinal directions and chaotic conditions for the intermediate directions.

[5] Cf. e. g., Pauly-Wissowa, RE VIII A, 2 col. 2325 ff.

[6] Dillmann, Lex. 180 and 791, respectively.

[7] BN 160 54ᵇ II, 17—19.

[8] BN 160 55ᵃ II, 3—4.

[9] Flemming, Henoch, p. 102, 10; Flemming-Radermacher, Henoch, p. 99. This gloss is simply omitted by Charles.

Passover

See Feasts (p. 142).

Philip of Opus

See Greeks (p. 165); Shadow Tables (p. 210).

Planets

In Ethiopic texts the planets appear exclusively under names transliterated from the Arabic (or even Latin)[1]. In the same context are also

[1] E. g., BM 816 7ᵇ to 9ᵃ. Grébaut, Patrol. Or. 6, p. 433 translates "des sept planètes du ciel" that which is actually "the 7 windows (masākwihā) of the heavens" (7 is probably a misreading of 6); cf. Book of Enoch 72, 7.

found rounded synodic periods, mean motions (often badly garbled)[2] and arrangements in concentric "rings" (śaragalā) or "heavens", remnants of an obsolete medieval tradition.

Astrological doctrines also reached Ethiopia through Arabic sources. E. g., the seasons and the winds are associated with the planets[3]:

Spring	humid-hot[4]	wind:	S	planets: ♃ and ♀
Summer	hot-dry		E	♂ and ⊙
Autumn	dry-cold		N	♄
Winter	cold-humid		W	☾ and ☿

Ethiopic names of the days of the week are taken from numerals[5] but eventually the planetary week also appears, based on the doctrine of planetary rulership over the hours[6].

In all texts of this type the planets are enumerated in the ordinary (Greek) sequence from Saturn to the moon. Only in the Slavonic Enoch[7] do we find the peculiar arrangement[8]

♄ ♀ ♂ ⊙ ♃ ☿ ☾ smallest stars

using Greek names, which is not surprising in a work based on a Byzantine original.

Another planetary sequence of unknown origin is attested in the following short section on astrological geography:

♄	Ethiopia, Nubia, . . ? . . , Middle-India
☿	Asia (?), Egypt
♃	Persia
♂	Country of the Tatars, Armenia
☾	Land of the Franks, Rome
♀	Yemen
⊙	Baghdad, Middle-India.

The terminology is Arabic[9].

---

[2] Vat 119 57ᵃ II, 20 to 57ᵇ I, 3 mentions direct and retrograde motion of Saturn.

[3] Vat 54 88ᵃ to 89ᵃ.

[4] The same pattern is also found in the "Golden Legend", ch. 33 (13th century).

[5] Excepting "Sabbath".

[6] Vat 128 103ᵇ II, 1 to 104ᵃ 2.

[7] Ch. XXX (Vaillant, p. 99/101) of the longer version; cf. s. v. Enoch, p. 113.

[8] Perhaps it is a simple mistake that interchanged the places of Venus and Jupiter.

[9] Berol 84 20ᵇ; BM 816 6ᵇ upper marg.; BN 160 9ᵇ, 21—30 (ignored in Zotenberg, Cat.).

Regulares

See Calendar, Appendix II (p. 65).

Rome

In our texts "Rome" appears in a variety of ways. One manuscript says[1] that a certain section on "hours" was written in Rome "in the days of King Qalamiṭos of Rome and of my King of Ethiopia Ja'qob"[2]. King Qalamiṭos is Pope Clement VIII (1592—1605) and Ja'qob ruled from 1597 to 1603 (and again in 1607)[3].

Another manuscript was sold "to Abba Justus, the Roman", in the 426th lunar cycle (i. e.) the "8th epact cycle"[4]. Since $426 = 8 \bmod 19$ the "epact cycle" is the 19-year cycle and 426 is the cycle number k in the 532-year cycle. The year in question is therefore $G_3\ 426 = J\ 1842 =$ A. D. 1849/50.

The julian calendar is another point of contact with Rome. Lists of parallel dates and names in different calendars are not rare; e. g., in the form of giving the Ethiopic dates for the first days of the julian calendar[5]

Sept. 1 = I 4
Oct. 1 = II 4
Nov. 1 = III 5

etc. which reflect the proper lengths: September = 30 days, October 31, etc. The names, however, that are associated with different calendars are frequently in disorder, e. g., when the Latin month-names are also given to a "Syriac" calendar so that "Syrian" December is the equivalent of "Roman" April[6].

There exist also some "Roman Computus" which somehow seems to have to do with non-canonical eras. In one case the "Roman Computus" seems to be based on an era which begins in W 38[7]. Another

---

1 Ambros B 22 fol. 164a, 8 to 11.

2 Cf. Galbiati [1957], p. 342.

3 Cf. Chaine, Chron., p. 246.

4 Vindob 6 18a (bottom).

5 E. g., Princ 5884 7b III, 2 to 10; Vindob 6 1a, 1 to 2a, 21; 3a, 10 to 15. Cf. also s. v. Length of Daylight, p. 182.

6 Princ 5884 12a II, 12 to 12b II, 3.

7 Dillmann, B. M. Catal. No. 11, p. 10 from BM Add 16220 5b II, 2—3.

"Roman Computus" gives a rule for finding the epact of a year from the Era Alexander[8]. Unfortunately the text is badly garbled and can only be tentatively reconstructed; under all circumstances it is not clear what Rome has to do with the Era Alexander.

Without visible motivation appears a rėmark beneath one of the standard calendaric tables: "from Rome and from Armenia"[9].

---

[8] Berol 84 5ᵇ I, 20—28; EMML 215 35ᵇ II, 7—24; 69ᵃ I, 4—16. The relationship of the Spanish era to the Christian era (Sp. 1 = B.C. 38) may well be parallel to the derivation of W′ from W (cf. above p. 126).

[9] Cf. s. v. Armenia (p. 19).

Saltus Lunae

See Maṭqeʿe and Epact (p. 189).

Scaliger

Joseph Justus Scaliger (1540—1609).

Ethiopian monks had a house of their own (Santo Stefano dei Mori) near St. Peter's in Rome from 1539 onward. Printing in Ethiopic type had begun in 1513 and an Ethiopic Grammar as well as an Ethiopic version of the New Testament were published in 1548/9[1]. Scaliger visited Rome in 1565 and 1566[2] and, with the help of the monks of San Stefano, gained a clear understanding of the Ethiopic calendar. He incorporated his results into his famous "De emendatione temporum" where he gave an excellent summary of the "Computus Ecclesiae Aethiopicae"[3].

Kepler said about Scaliger "cuius multiplicem eruditionem iure merito mirantur omnes, imitabuntur paucissimi"[4].

---

[1] Cf. Ullendorff, Ethiop., p. 8.

[2] Bernays, Scal., p. 39 and p. 131, note 15. Cf. also Vat. Catal. II, p. 16: "Domus Vaticana s. Stephani, quae saec. XVI tot exceperat hospites."

[3] I am quoting from the edition of 1629 (pp. 670 to 706). I collated the printed Ethiopic text with the original manuscript, now in Leyden. Weld Blundell (Roy. Chron. p. 494) stated incorrectly that BM Add 16217 was Scaliger's source; the source of error is evidently a note by Dillmann (BM Cat. p. 45, note (a)) where he refers to the similarity between the rules in BM Add 16217, fols. 10—18, and Scaliger's text.

[4] Astronomiae pars optica (1604) VI, 11 (Werke 2, p. 225).

Seasons

See also: Winds (p. 228).

It is evident—e. g., from the schemes for the length of daylight, from the shadow tables, or from the positions of the sun at sunrise with respect to the "gates"—that the Ethiopic seasons are based on a four-division of the year, more or less in agreement with the division by the equinoxes and solstices. This type of definition reflects the Babylonian-Greek astronomical tradition, in contrast to the three agricultural Egyptian seasons of four months each.

The actual definitions, however, are not based on astronomical considerations but on simple arithmetical schemes. We know of two (strictly speaking, contradictory) traditions which, however, are always intertwined in our sources. One follows the rigid pattern of Enoch's calendar: four seasons of 91 days each, with months counted from Spring[1]; the other is based on the Ethiopian (i. e. Alexandrian) civil calendar, with New Years (accidentally) near Autumn and a fixed date, 25, for the equinoxes and solstices. In this second case the "seasons" are actually of unequal length because of the epagomenal days: three seasons are 90 days long, the fourth contains 95 (or 96) days.

In the year of Enoch the months in each season have 30, 30, and 31 days, respectively, and are counted as follows[2]:

(1)	Spring:	months 1 to 3
	Summer:	4 to 6
	Autumn:	7 to 9
	Winter:	10 to 12.

One can also speak about four "corners" (ma'azen) of 13 weeks each[3] or simply 4 quarters of 91 days[4]. In the Ethiopic civil calendar the seasons are located as follows[5]:

Autumn:	I 26 to IV 25
Winter:	IV 26 to VII 25
Spring:	VII 26 to X 25
Summer:	X 26 to I 25.

---

1 Book of Enoch, ch. 72, 3 to 36; cf. also ch. 82, 4, 6, 11 to 19.

2 Berol 84 32ᵇ II, 4 to 17; also 27ᵇ II, 11 to 28ᵃ III, 13.

3 Berol 84 5ᵃ I, 3 to 5.

4 Vindob 6 32ᵇ I, 12 to 15.

5 Berol 84 4ᵇ I, 15 to II, 7; 12ᵇ II, 4 to 12; 28ᵇ II, 10 to 29ᵃ I, 1; 41ᵇ I, 1 to II, 20; Vindob 6 32ᵇ II, 16 to 33ᵃ I. 15.

The date for the equinoxes is 25 (not 26), as we are explicitly told[6].

The origin of this norm seems to be unknown[7]. The fact, however, that VII 25 is the earliest date for Passover[8] suggests a derivation from the pattern of the Jewish 19-year cycle in its relation to the Alexandrian civil calendar.

The influence which the Book of Enoch exercised on much later periods is strikingly illustrated by a section in one of our treatises[9] which pretends to determine accurately the moments of beginning for each season. The underlying method, however, has nothing to do with astronomy but assumes dogmatically a length of 91 days for each season. Otherwise the "year" of the author is the Ethiopic civil year of 365 days, so that the four "seasons" are one day shorter than one year. Consequently each single season recedes 1/4 of one day. The text therefore begins with a four-division of the day (at the first and seventh hour of daytime and night, respectively) and then tells us

if spring begins in the first part of the day
then summer in the 4th, autumn in the 3rd, winter in the second;

(3) if spring begins in the second part of the day
then summer in the 1st, autumn in the 4th, winter in the third; etc.

This shows that the discrepancy between the two calendar systems was not only not eliminated but carefully canonised[10], without ever considering the long-run consequences.

The coexistence of two contradictory definitions gave rise to terminological ambiguities which are apt to obscure the meaning of some sentences. As we have remarked, the months of Enoch have no names but only numbers which begin at the vernal equinox, which is identified

---

[6] E. g., CSCO 221, p. 115 1 f./222 p. 100 with reference to the Didascalia and to VII 26 as lower limit for Easter; similarly Berol 84 29[a] I, 1 to 6; 41[b] I, 25 to 27.

[7] Cf. the Alexandrian solstice date Payni (X) 25, quoted by Cosmas Indicopleustes, about A. D. 550 (Top. Chrét. VI, 2, ed. Wolska-Conus III p. 14/15). I know, however, of no other evidence for exactly this norm.—The julian equivalent of VII 25 is March 21.

[8] Cf. s. v. Feasts p. 133.

[9] Berol 84 32[b] I, 11—34[a] I, 14.

[10] This, by the way, refutes explicitly all modern theories concerning intercalations in the Enoch calendar.

with month 1 day 1. An Ethiopic text may use two ways in referring to these months: either by numerals (which is correct) or by calling month 1 again "Maskaram" as if it were month I of the Ethiopic calendar. Therefore "Miyāzyā" (VIII) can also mean month 8 in the Enoch calendar and a sentence[11] "the month Miyāzyā begins on the sixth of Miyāzyā and ends on the fifth of Genbot" means that a civil month begins on the 6th of month 8 and ends on the 5th of month 9 (cf. Fig. 5). In the civil

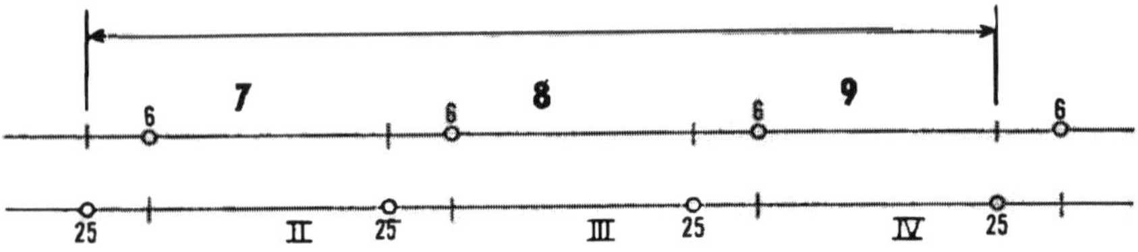

Fig. 5

calendar its name would be Ḫedar (III) but from the viewpoint of the Enoch calendar it is also named Miyāzyā. The interval of 6 days separating the two calendars (which are ignored in the naming of the months), originates in the assumption that the months of Enoch begin in principle at the vernal equinox to which is assigned the date 25 in the civil calendar. In the same way the civil month "VIII" (from VII 26 to VIII 25) corresponds to Enoch month 1 (cf. Fig. 6), so that[12] "these three months

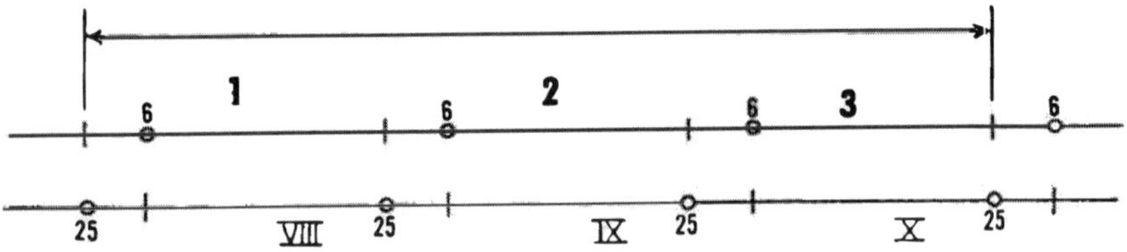

Fig. 6

(VIII, IX, and X) are the time of spring for the Hebrews, but the time of autumn for Ethiopia". In other words "VIII, IX, and X" can be interpreted in two ways: either in the Ethiopic calendar, i. e. spring; or as 8, 9, and 10 in the Enoch calendar, i. e. essentially autumn[13].

---

11 Vat 106 91^b I, 19 to 21.

12 Vat 106 91^b II, 9 to 11.

13 There remain other passages which I do not understand. E. g., Magābit (VII) 25, i. e. the vernal equinox, is said to be "Magābit 22 according to

It is of interest that here the "Hebrews" are the adherents (or inventors) of the calendar which is otherwise associated with the name of Enoch.

The ambiguity in the use of Ethiopic month-names is found also in other contexts. For example, four intervals of 91 days each are associated with the seasons as follows in a treatise called "Computus of the Branch"[14]:

(4)	from IV 30,	91 days:	summer
	from VIII 1,	91 days:	autumn
	from XI 2,	91 days:	winter
	from I 28,	91 days to IV 28:	spring.

Here the Ethiopic month-names cannot refer to the civil calendar in which I is near the beginning of autumn. Hence we must be dealing with the months of Enoch, although one should expect I 1 instead of I 28 as the beginning of spring. The one day, IV 29, that is missing to complete in (4) the civil year is reserved for the birth of Christ[15].

As a result we can say that the seasons are arranged as follows:

(5)	autumn (ḍaday):	II to IV	7 to 9
	winter (keramet):	V to VII	10 to 12
	spring (maḍaw):	VIII to X	1 to 3
	summer (ḥagay):	XI to I	4 to 6.

The ambiguities are caused by the difference in numbering of the two systems, never by the commonly used names of the seasons, listed in (5).

Another set of names exists, however, which were taken from the Book of Enoch[16] where the four lead stars that separate the seasons are given these names[17]. In this way one obtained the following terminology:

---

the Hebrews" (CSCO 221, p. 115, 10/11; 222, p. 101; CSCO 235, p. 102, 21/22; 236, p. 90; also Didascalia V, 17 Harden p. 131).

[14] Upps 3 62ᵇ 4 to 14; similarly, BN Abb 37 4ᵃ, lower margin 1—4. Löfgren's suggestion (Katal. p. 30) that "Branch" means here "season" is very implausible in view of the next section which is called "Computus of the Trunk". There are (line 15) 12 "branches" of 30 (days) each, the "trunk" being the year.

[15] This is reminiscent of the Egyptian mythology which makes the epagomenal days the birthdays for the children of the sky goddess Nut.

[16] Ch. 82, 13. Also Vat 106 91ᵇ II, 11/12; 92ᵃ I, 16; II, 19; 92ᵇ I, 18; Vat 119 66ᵇ I, 18 to 69ᵃ I, 10.

[17] Cf. s. v. Stars, p. 216.

	autumn:	mēl'ēyal
(6)	winter:	nārēl
	spring:	melke'ēl
	summer:	hel'emmēlēk.

The scribes frequently misspelled these unusual personal names[18] and it is therefore not always clear which season is meant.

The concept of "seasons" was even extended to the lunar year of 354 days that underlies the pattern of the "gates"[19]. There, 12 months are introduced, beginning with "Miyāzyā (VIII) called Nisan", which are alternatingly full and hollow. All these months, with the usual Ethiopic names, begin on day 6 of month M and end on the fifth of month M + 1. Each triple of months (beginning with VIII, IX, X) is one season (alternating 89 and 88 days, thus 177 days for six months, as the text explicitly says). These seasons are called

"in Hebrew"	maṣaw,	"in Ethiopic"	ḍaday,	"star"	melkā'el
	ḥagāy		keramet		ḥelmalēk
	ṣaday		maḍaw		mēl'iyal
	keramet		ḥagāy		nāhēl.

It is clear that we have here another instance where years and seasons can be counted from diametrically opposite points. But I do not know why day 6 was chosen as a starting point. Epagomenal days do not concern this schematic lunar year.

In a text written under Islamic influence, we find dates for the solstices preceding the ordinary norm by 8 days (Thebet 17 and Tammuz 17 respectively)[20], assuming that the Syrian months are co-terminal with the Ethiopic months IV and X. The dates for the lunar mansions suggest VIII 23 as vernal equinox[21].

Finally, we must mention a theological postulate for the date of the (autumnal) equinox that disagrees with all norms otherwise used for the seasons. As we know, the 532-year tables are based on the assumption that the calendar began on Tuesday, the day of the creation of the moon.

---

18 E. g. mikā'el (i. e. Michael) for autumn (Berol 84 24ᵇ, 4; Upps 3 73ᵇ, 2) or ḥelmomamā'ēl for summer (BM 816 16ᵃ). Cf. also the apparatus in Flemming, Henoch, p. 113.

19 From EMML 215 37ᵃ I, 12—39ᵃ I, 23, with many scribal errors in the day numbers, but easily emended.

20 Vat 54 88ᵃ, 9 to 89ᵃ, 19, and fragmentary parallels in Ambros B 22; cf. also s. v. Enoch, Appendix p. 113, ch. XLVIII.

21 Cf. s. v. Lunar Mansions, p. 183.

Consequently the tables give for W 1, Maskaram 1 the weekday $t = 7$. For this day it is then assumed that both daytime and night are 12 hours in length[22], i. e. that I 1 is the date of the equinox. I know of no consequences derived from this definition.

[22] BM 815 22ᵇ II, 11—15 and parallels (e. g. BM Add 24995 26ᵃ II, 10—15; Vat 119 89ᵃ II, 12—17).

Shadow Tables

1. Summary

I know of well over thirty Ethiopic shadow tables. They are all based on the same principle: the length of the shadow, measured in feet, of a man standing upright, defines the hour according to a simple arithmetical pattern. These hours are "seasonal hours" because their total is always 12. This fact alone is indicative for a hellenistic origin[1].

The increments of the shadow lengths during the hours before or after noon are always 1, 2, 3, 4, and 10 feet. The noon shadow, however, from which these differences are reckoned, depends on the months, assuming again a simple arithmetical sequence: they vary linearly, either with a one foot monthly difference, between $u = 2$ and $U = 8$, or with a difference of $1\frac{1}{2}$ between $u = 0$ and $U = 9$. In the first case the extrema usually lie in the months XI and V, in the second case in X and IV.

It is obvious that these tables are valueless for measuring time in Ethiopia. At a geographical latitude of, e. g., $\varphi = 14°$ (Axum), the noon shadow will be zero twice in each year and vary for a gnomon of length 6 between about 5 to the north and 1 to the south (cf. Fig. 7). That a man could cast a noon shadow of 8 or 9 feet is out of the question. Conversely, a shortest noon shadow 0 requires a geographical latitude of about 24° (Syene), whereas a longest shadow of about 8 feet would agree with about $\varphi = 38°$ (Athens). Limits like $u = 2$, $U = 8$ are theoretically incompatible and obviously chosen for the simplicity of the linear variation of the basic numbers. At any rate it is obvious that these tables can only have originated in the mediterranean area and were taken over by the Ethiopians without the slightest understanding.

In many of our texts the shadow tables proper are accompanied by an explanatory text which describes the arithmetical structure of the

[1] Cf. also s. v. Hours, p. 167.

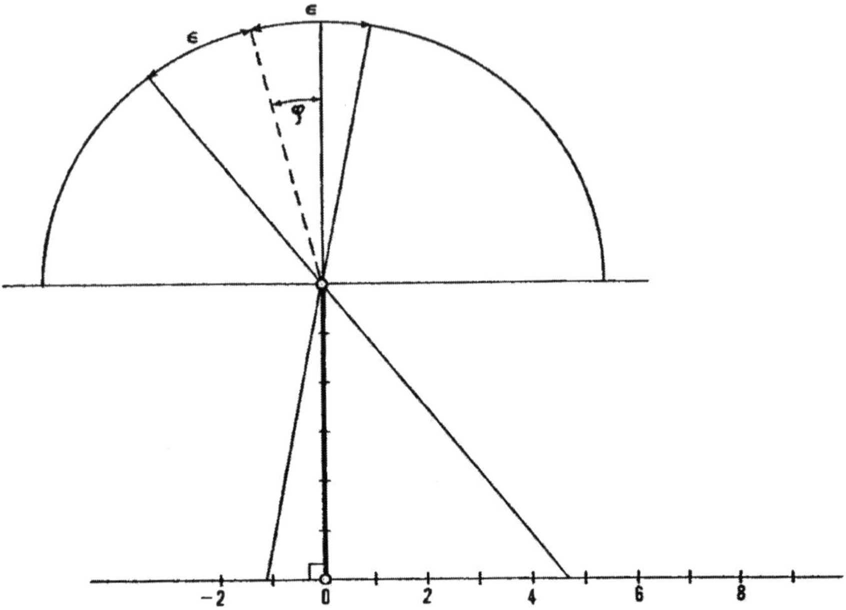

Fig. 7

tables. All of these texts which concern tables with the extrema $u = 2$, $U = 8$ name as their authors "Synesius and Philip, kings of the Greeks"[2]. Of course, these men are not "kings", but otherwise well known for their astronomical interests. Synesius (around A. D. 400) left us a treatise on the astrolabe, and Philip of Opus's astronomical data (from the middle of the 4th century B. C.) are known to us through Ptolemy's "Phaseis". Indeed we have Greek shadow tables, based on the same parameters, one from the Ptolemaic period, others (Byzantine) dedicated to "King Philip"[3].

On the other hand, tables and explanatory texts based on the extremal noon shadows $u = 0$, $U = 9$ usually do not mention authorities or they speak about "hours of the Copts" (gebṣ) which suggests a later (Alexandrian) version of the earlier Greek tables.

Many more shadow tables, all equally useless, are preserved from the Middle Ages, from western Europe as well as from Syria and Armenia, copied in the same mechanical fashion as in Ethiopia[4]. All these tables

[2] Cf. s. v. Greeks, p. 165 n. 5.
[3] Neugebauer [1962] p. 32—37.
[4] Cf. Neugebauer, HAMA II p. 736 ff.

are eventually derived from a very early and primitive phase of Greek astronomy.

The use of the Ethiopic shadow tables is extended even to some hours of the night[5], assuming that one enters the tables for a month six months later than for daytime[6]. No explanation is given for this peculiar procedure.

It is not surprising that one cannot find in the Ethiopic literature evidence for the use of our tables. But even if we find some shadow lengths mentioned, there is trouble. For example, in the Annals of King ʿĀlam Sagad (A. D. 1667) we are told that the shadow after completion of the third hour in Ṭeqemt (II) 8 was three feet, but only three days later at the same hour four feet[7]. Obviously no accuracy can be expected from such data.

2. Classification of the Tables.

In common to all tables is the pattern of increments of shadow lengths in relation to the noon shadow:

(1a) 1 2 3 4 10 feet.

That is to say, if u_i is the length of the shadow at the i-th hour, then

(1b)
$$u_5 = u_7 = u_6 + 1$$
$$u_4 = u_8 = u_5 + 2 = u_7 + 2$$
$$\text{etc.}$$
$$u_1 = u_{11} = u_2 + 10 = u_{10} + 10 = u_6 + 20.$$

In all tables one finds

(1c)
$$u_{12} = 70^f$$

throughout the whole year. This number is the result of a misinterpretation of a Greek zero-sign (meaning "no shadow" at the 12th hour) as an omicron, i. e. as the numeral sign for 70.

The main distinction between the different tables is based on the values of the noon shadows for the different months. Tables with the extremal noon shadows $U = 8$, $u = 2$ use the following linear sequence

(2a) 8 7 6 5 4 3 2 feet

for the months

[5] BM 816 44ª I, 18—23; parallel, Vindob 6 22ᵇ I, 14—18 (garbled).

[6] Vindob 6 22ᵇ I, 18—II, 22. The epagomenal days are included with month XII (Vindob 6 22ᵇ I, 18; BN Abb 37 17ᵇ, margin).

[7] CSCO 22, p. 4, 2 f. and 19 f.; 23 p. 2.

		VI	VII	VIII	IX	X	
(2b)	V						XI
		IV	III	II	I	XII	

whereas $U = 9$, $u = 0$ lead to the noon shadows

(3a) 9 7½ 6 4½ 3 1½ 0 feet

for the months

		V	VI	VII	VIII	IX	
(3b)	IV						X
		III	II	I	XII	XI	

Table I

hour	IV	V / III	VI / II	VII / I	VIII / XII	IX / XI	X
1	29	27½	26	24½	23	21½	20
2	19	17½	16	14½	13	11½	10
3	15	13½	12	10½	9	7½	6
4	12	10½	9	7½	6	4½	3
5	10	8½	7	5½	4	2½	1
6	9	7½	6	4½	3	1½	0
7	10	8½	7	5½	4	2½	1
8	12	10½	9	7½	6	4½	3
9	15	13½	12	10½	9	7½	6
10	19	17½	16	14½	13	11½	10
11	29	27½	26	24½	23	21½	20
12	70	70	70	70	70	70	70

The complete table derived from the pattern (3a) and (3b) is shown in the subsequent Table I. Several copies of it, or of a slightly modified type, are extant. I do not know, however, of a corresponding 7-column table for parameters (2a) and (2b). All tables based on $U = 8$, $u = 2$ (and many tables with $U = 9$, $u = 0$) display their numbers in 12 columns.

Many of these tables are in bad disorder, the main cause being a contraction of a 7-column table into a table with 6 pairs of months.

Such a contraction necessitated the elimination of a whole column of numbers and thus to a disturbance of the arithmetical structure at some arbitrary place. This then became the source of additional disorder. Several tables, based on the sequence (3a), simply drop the halves after 1, 4, and 7 (and, of course, in the rest of these columns).

One may think that the 12-column arrangement is the original one. But we have clear evidence that 12 column tables were also created by unfolding of the 7- or of the 6-column type. In the latter case the nissing numbers had to be replaced somehow, e. g., by simply duplicating another column (a well known trick of scribes to fill in a gap). The final result in almost all tables is a huge number of errors which alone would suffice to make these tables useless.

Scribal errors not beeing enough, some tables were derived from Table I by regrouping the columns as follows (Table II) for reasons unknown.

Table II

I	II	III	IV	VIII	IX	X
VII	VI	V		XII	XI	
4½	6	7½	9	3	1½	0
	→				←	

or

VIII	IX	X	IV	I	II	III
XII	XI			VII	VI	V
3	1½	0	9	4½	6	7½
	←				→	

The two main types of shadow tables differ not only in the values of U and u but they correspond also to different positions of the equinoxes and solstices (Table III).

Table III

	$U = 8$, $u = 2$	$U = 9$, $u = 0$
autumn equinox	in II	in I
winter solstice	V	IV
spring equinox	VIII	VII
summer solstice	XI	X

The first norm, equinoxes in II and VIII, is more in accord with the usual Ethiopic division of the seasons at I 25, IV 25, etc.[8], than is the second norm which is, however, the only one which is also used in the schemes that describe the variable length of daylight[9].

It is difficult to understand how one could introduce the norm $u = 0$, $U = 9$. Perhaps one motive was the desire to include the south-

[8] Cf. s. v. Seasons, p. 204.

[9] Cf. s. v. Length of Daylight, p. 180.

ern limit of Egypt, maintaining at the same time the traditional mean value which corresponds to latitudes of about $\varphi = 36°$ or $38°$. Combined with the value $u = 0$ at Syene, a constant difference of $1\frac{1}{2}$ leads then to $U = 9$. Such a procedure sounds rather inept but would not be out of lines with other arithmetical patterns.

3. *Textual Evidence*

I. Tables

A. $U = 8, u = 2$

$U = 8$ in V, $u = 2$ in XI; 12 columns	
beginning with V:	beginning with I:
Berol 84 16[a, b]	EMML 941; 2114
BN 50 3[b]	Princ 5884 14[a] I/II
EMML 2077 5[b]—6[b]	
Vat 119 56[b]	
Vat 128 105[a]	
$U = 8$ in IV, $u = 2$ in X; 12 columns	
beginning with VII:	
BM Add 16217 9[b]	

B. $U = 9, u = 0$

Arrangement as in Table I:	as in Table II, left:	as in Table II, right:
EMML 640; 3495	Berol 84 15[b]	BM 816 43[a] lower half
Upps 3 76[a, b]	Upps 42 95[a]	EMML 49 29[a] lower half
Vat 123 66[b]	Vat 171 91[b]	EMML 160 275[a] lower half
		EMML 215 33[a] lower half
12 columns, beginning with I:	beginning with IV:	beginning with V:
Ṭānā 83 103[b]	BM 816 43[a] upper half	BN Abb 37 17[b]
	EMML 49 29[a] upper half	
	EMML 160 275[a] upper half	
	EMML 25 33[a] upper half	
	Vindob 6 21[b]	

C. Hybrid Type: $U = 7, u = 1$ for 4^h to 8^h, $U = 6, u = 0$ for other hours
12 columns, beginning with I, U in IV, u in X: BM 827 118[a]

II. Explanatory Texts

A. $U = 8, u = 2$ ("Synesius and Philip")

U in V, u in XI:	U in IV, u in X:
Berol 84 15^{b} I, 1—16^{a} II, 11	
BN 50 3^{b}, 1—31	
EMML 215 66^{a} I, 21—II, 26	EMML 215 72^{a} I, 27—72^{b} I, 19
Vat 119 56^{a} I, 1—57^{a} I, 5	
Vat 128 104^{b} I, 1—II, 16	

B. $U = 9, u = 0$

U in V, u in XI:	U in IV, u in X:
	BM 816 43^{b} I, 1—44^{b} I, 23
	BM Add 16217 10^{a} I, 1—10^{b} II, 17
	BM Add 24995 29^{b} I, 20—30^{b} I, 21
	EMML 215 70^{a} I, 12—70^{b} II, 26
	EMML 2077 154^{a} I, 1—154^{b} I, 16
	Jerus Arm 3482 1^{b} I, 1—3^{b} I, 20
Vindob 6 22^{a} I, 1—22^{b} I, 9	Vindob 6 22^{b} I, 10—13

Sibyls

The Jewish feast "Inner of Earth" is related to the "Erythrean" Sibyl[1]. A "Computus of the Sibyl" divides the 7000 years of the World into nine "suns", or nine "ages", concepts also associated with the Tiburtine Sibyl[2]. The occurrence in the Ethiopic material suggests borrowing from earlier Jewish material.

[1] Cf. s. v. Feasts, p. 143.
[2] Cf. s. v. Jubilees, p. 174.

Slavonic Enoch

See Enoch, Appendix (p. 111).

Solar Cycle

See Cycles (p. 87).

Solar Epact

See Ṭentyon (p. 219).

Solstices

See Seasons (p. 204).

Stars

Nowhere in our material can one find references to constellations (zodiacal or otherwise) or to single fixed stars or planets except in texts that depend on Arabic sources. This does not mean, however, that "stars" are not mentioned in Ethiopic works; quite to the contrary in all descriptions of the different heavens which were shown to Enoch an abundance of stars served as celestial decoration, yet without any relation to astronomy. In fact these "stars" only reflect a more or less military-like organization of the angels. The four great "guiding stars" that divide the year into four 91-day seasons carry personal names just like the angels (e. g. ending in -ēl)[1]. There are furthermore 1000 stars for the remaining 360 days and 12 stars for the 12 months[2]. In the Slavonic Enoch the four great stars seem to have been duplicated because four accompany the sun at the right as well as to the left, each commanding 1000 stars[3].

In the original composition of the Book of Enoch the stars are related to the pattern of the six "gates" through which sun and moon rise[4]. Three "splendid stars" are associated with the gates of each season, in the order

(1a) 4 5 6; 6 5 4; 3 2 1; 1 2 3

which corresponds to the order of the Ethiopic months,

(1b) VIII IX X; XI XII I; II III IV; V VI VII

i. e. to the sequence of the four seasons[5]. This is exactly the order given for the conjunctions of sun and moon from month to month[6]. Finally the four lead-stars belong to the gates 4 6 3 1, as is appropriate to their role as dividers of the seasons[7].

---

[1] Cf. s. v. Seasons, p. 208.

[2] Enoch, ch. 82. Similarly in other treatises, e. g. Berol 84 32^a III, 21 to 32^b I, 10; 33^a II, 5 to 19.

[3] Bonwetsch [1922] p. 11, ch. XI, 3; Vaillant, Secrets, p. 13; p. 91.

[4] Cf. Vat 106 91^b I, 7 to 92^b I, 21.

[5] Cf. s. v. Seasons, p. 204.

[6] Cf. s. v. Gates, p. 158

It seems rather evident that we have thus far circumscribed the whole "astronomical" contents of the Book of Enoch: the 4-division of the year of 364 days and the pattern of the gates for the solar year and the schematic lunar year.

In later Ethiopic material, however, we find some faint traces of a slightly more advanced knowledge, e. g., a short discussion about the periods of invisibility of stars in the course of the year, though in a qualitative fashion only[8]. Then we have evidence for a distinction between sidereal and solar time resulting in a "delay" of two hours each month, thus of one day in one year[9]. It is certainly not surprising that such concepts eventually reached Ethiopia, having been common knowledge in Greek astronomy for centuries.

A different type of influence is due to the Islamic penetration of Ethiopia. For example, the year of 364 days is now associated with "28 stars" for 13 days each, an obvious adaptation of the concept of the lunar mansions[10]. Even extensive excerpts from Ptolemy's catalogue of stars in the Almagest are found in Ethiopic[11], in particular the summary of the distribution of stars of different magnitudes within the total of 1022. A list of the apparent diameters of fixed stars from the first to the sixth magnitude begins with 108 and decreases linearly with the difference of 18 parts ('ede) to 18 for a star of sixth magnitude, where a "part" probably represents a second of arc. I do not know of these parameters in Greek astronomy.

Appendix

I know of no Ethiopic parallel of an Aramaic fragment on "stars" published by Milik[12]. Unfortunately the text is so badly preserved that one cannot make any consistent sense of it. There are fractions of sixths and ninths and 1, 2, and 3 days (?). In principle one could expect something about "windows" in relation to the risings of the lead-stars.

---

[8] Berol 84 33^b II, 21 to III, 13. I do not understand the details and I suspect neither did the scribe.

[9] Berol 84 33^b I, 12 to 15; cf. also s. v. Year, p. 230.

[10] Vindob 6 32^b I, 16 to 18.

[11] BM 816 10^a I, 24 to II, 12.

[12] Milik, Enoch, p. 296 f. What he means by "the celestial sphere divided into 360 degrees" or by "the astral evolutions" I do not know.

Sun

See Moon (p. 194).

Sunday

Called "Christian Sabbath" or "First (Day)" of the week[1]. The liturgical association of some Christian feasts with Sunday created an interest also in the weekdays of Jewish feasts for the purpose of avoiding coincidences. Hence we have in the calendaric tables a special column (*pw*) for the weekday of Passover and statements about the frequency of Passover falling on a Sunday[2].

On the other hand, we find rules in our texts about how to determine the civil dates of the Sundays in a year of given ṭentyon[3].

[1] For the general background cf. E. Hammerschmidt, Stellung und Bedeutung der Sabbats in Äthiopien (Stuttgart, 1963).

[2] Cf. s. v. Calendar, p. 52; Feasts p. 150.

[3] Cf. s. v. Week, p. 225.

Sundial

See Shadow Tables (p. 209).

Synesius of Cyrene

See Greeks (p. 165); see Shadow Tables (p. 210).

Syriac

See Months (p. 194).

Tables

See Calendar, Tables (p. 29); Shadow Tables (p. 211).

Ṭentyon and Day of John

The term ṭentyon (*t*) denotes the weekday of the first day of the Ethiopic year[1], Maskaram 1, counting Wednesday as 1, Thursday as 2, etc. Its equivalent, counting Sunday as 1, is called "Day of John" (*j*). Both these norms can be directly established by comparing the numbers in the columns *j* and *t* of the 532-year tables with the explicit names of

[1] The interpretation as "Sunday-Letter" (Rüppell, Reise, II p. 38; Ginzel, Hdb. III, p. 326) is wrong.

the weekdays in column ε^2. The connection with the first day of the year is expressed by the variant term for column j, "Beginning of Thoth" (śarqatito)[3], the month Thoth being the exact equivalent of Maskaram. Also all computational rules involving the parameters t or j agree with the meaning "weekday of I, 1".

In spite of this clear evidence modern scholars have accepted a derivation of the term ṭentyon from Greek πλινθίον, suggested by Scaliger[4] in 1583 and incorporated into his Lexicon by Dillmann. In fact, however, nowhere in Greek astronomy is πλινθίον attested in a calendaric context. Only a certain type of sundial is described by this name in Vitruvius[5], a fact that caused flagrantly impossible translations for ṭentyon[6]. The Ethiopians themselves, being familiar with the use of t in the 532-year tables, gave at least a factually possible explanation by relating ṭentyon to ṭenta, "beginning"[7].

Yet the correct derivation is not difficult to find[8]. Paulus Alexandrinus (about A. D. 375), in a passage (referred to by Scaliger!), gives essentially the same rule as the Ethiopic texts for finding the weekday of a given day[9]. There he uses for weekday the expression τῶν θεῶν, i. e. "(day) of the (planetary) gods", a terminology obviously commonly known since Athanasius also used it in his Easter messages (328 to 373)[10]. Thus ṭentyon, or ṭentewon[11], is simply a loan-word from the Greek, just as 'abaqtē, or paguemēn, etc.

An equivalent name for the parameter t is "solar epact"[12], chosen in analogy to "lunar epact" (m) which represents the first day of the

---

[2] Cf. s. v. Calendar, Table III, p. 31.

[3] Cf. below, p. 222.

[4] De emend. temp., p. 691 and p. 688 in the edition of 1629.

[5] De archit., Book 9, VIII 1.

[6] Conti Rossini: "(heur du) cadran solaire" (CSCO 58, p. 43, 14/15; p. 44, 20). Abbadie: "sorte de cadran solaire" (Catal., p. 107, n. 1). Mauro da Leonessa [1943] even quotes one translation as "meridian" (p. 319, n. 1).

[7] E. g. Berol 84 3a I, 11 to 14; Upps 3 70a, 3 to 5; cf. also Dillmann, Lex. 942 and 1392; Chaine, Chron., p. 107. Of modern scholars only Huntingford (Charters, p. 48, n. 1) accepted the native etymology.

[8] Neugebauer [1976].

[9] Paulus Alex., Eisag., ed. Boer, p. 40 in an example for Diocletian 94 (A. D. 378).

[10] Schwartz, Ostert., p. 4; from Syriac (cf. Piper, Kal., p. 111 f.); also Schwartz [1904] p. 335, p. 338/9, n. 2 (= Ges. Schr. 3, p. 4, p. 7/8 n. 2).

[11] This spelling in Berol 84 1a II, 12; Upps 3 65a, 2, 10, et passim.

[12] 'Abaqtē ḍaḥay (e. g. Berol 84 15a II, 13; 39b II, 11/12). The "Guide for the Blind" (Guidi [1896] p. 377, II, 7—11) uses the term "solar ṭentyon".

lunar year[13]. The identity of "ṭentyon" and "solar epact" follows, e. g., from the use of identical computational rules[14]. A description of t as "beginning of the sun"[15] fits into the same framework, particularly since the sun was created on a Wednesday, i. e., on $t = 1$.

Even medieval chronology has preserved the same terminology. "Epactae solis" and "concurrentes septimanae" are equivalent terms, both referring to a parameter that has the identical numerical value as t[16]. The corresponding Greek term is πόσαι τῶν ϑεῶν[17].

I do not know of explicit textual evidence that would motivate the counting of Wednesday as $t = 1$. If one looks, however, at the weekdays of Maskaram 1 in the four Ethiopic eras[18] one finds

Table I

Year 1 of Era	Cycle No. k	Cycle No. c	Maskaram 1 Weekday	t_c
W	1	1	Tuesday	7
J	181	10	Wednesday	1
D	457	1	Friday	3
G	1	1	Tuesday	7

This suggests the Era of Incarnation as a starting point for the numbering of t.

The importance of the parameter t within the calendar-computus is made evident by the fact that the date of "Nineveh" (n) is uniquely determined by t and the cycle number c[19]. Thus t and c determine the dates of all Christian festivals in a given year.

---

[13] Cf. s. v. Maṭqeʿe and Epact, p. 187. It should be noted that the term "solar epact" is used by Abū Shāker in his Computus (ch. 23) for a parameter which has no relation to the ṭentyon (cf. above p. 17).

[14] Vat 106 (Vat Catal. p. 389, ch. 23/24).

[15] BM Add 24995 25ᵇ I, 2 and I, 15—17; Vat 119 88ᵃ I, 11/12.

[16] Cf. s. v. Calendar, Appendix II (p. 64).

[17] Rühl, Chron., p. 142 and p. 118; Ginzel, Hdb. III, p. 143 ff.

[18] Cf. s. v. Eras, Tables I and II (p. 117 f.).

[19] Cf. s. v. Calendar, Table XIII, p. 50.

Computation of t

The intervals in years between the four Ethiopian eras are all divisible by 4. Hence the leap years are the same for all eras, falling at the cycle years $k = 3, 7, 11, \ldots$ etc.

Since $365 \equiv 1 \bmod 7$ the value of t increases by 1 for each ordinary year and by 2 for a leap year. Consequently the years $k = 4, 8, 12, \ldots$ etc. have values of t which are 2 higher than for the preceding years (counting, as always, modulo 7). It is now easy to construct such a sequence by using the consecutive year numbers E of any one of our eras. If $[E/4]$ denotes the greatest integer contained in $E/4$ (thus $[E/4] = 1, 2, 3$, etc. for $E = 4$ to 7, 8 to 11, 12 to 15, etc.) and if t_0 is the value of t for the first year of the era, the following sequence has the desired quality to increase by 1 in ordinary years and by 2 in leap years:

$$\begin{array}{lll} \text{for } E = 1 & t = t_0 & = t_0 + E - 1 \\ 2 & t_0 + 1 & = t_0 + E - 1 \\ 3 & t_0 + 2 & = t_0 + E - 1 \\ 4 & t_0 + 3 & = t_0 + E - 1 + [E/4] \\ 5 & t_0 + 4 & = t_0 + E - 1 + [E/4] \end{array}$$

hence in general

(1) $$t \equiv t_0 + E - 1 + [E/4] \quad \text{mod } 7.$$

Table I (p. 220) shows that $t_0 \equiv 0 \bmod 7$ for the eras W and E, but 1 for J and 3 for D. Hence t can be found from either one of the following relations (mod 7):

(2a) $$t \equiv W + [W/4] - 1 \equiv G + [G/4] - 1$$

(2b) $$t \equiv D + [D/4] + 2$$

(2c) $$t \equiv J + [J/4].$$

These rules exactly are found in our treatises[20]. The additive constants, 2 in (2b) and -1 in (2a) bothered the Ethiopian chronologers who invented theological explanations based on the situation at creation[21]. In fact they result from the accident that Thoth 1 of Diocletian 1 was a Friday.

For the rule (2a) one usually finds the formulation

(2d) $$t \equiv W - 1 + \tfrac{1}{4}(W - 1) \quad \text{mod } 7.$$

Here, however, one cannot simply take the integer part $[1/4(W-1)]$ of

[20] Vat 119 88[a] I, 10 to II, 7 and Ṭānā 34 101[b] I, 16 to 25 (for W); Upps 3 65[a] 2 to 11, and Scaliger, p. 673, 17 to 21 (for G).

[21] E. g. Berol 84 3[a] I, 9 to 16; Upps 3 70[a], 1 to 8.

the fractional term because this would locate the increment 2 of t one year later than [W/4]. This was indeed realized by our computists since we are told[22]: "if the residue is 3/4 add (one), if it is less than 3/4 ignore it". This procedure produces the same results as (2a).

Day of John

The 532-year tables display, parallel to the column for the ṭentyon, a column headed "Day of John" (ʿelata yoḥanes)[23] that gives numbers (j) in fixed relation to t:

(3) $$j \equiv t + 3 \equiv t - 4 \mod 7.$$

Since $t = 1$ represents Wednesday, we have now $j = 1$ for Sunday. The function of this new parameter is therefore simply to change the norm of the ṭentyon to the ordinary count that starts with Sunday, i. e. to the "feria" of medieval terminology. The name "Day of John" originates from the feast of John the Baptist on the civil New Years Day[24]. The connection of j with the first day of the year is expressed in the equivalent term "Beginning of Thoth" (śarqatito)[25].

Another terminology for j is ḥasāba samāy wameder[25a], i.e. "Computus (or Count) of Heaven and Earth", probably in reference to the sequence of the weekdays at Creation.

For the computation of j it follows from (3) in combination with (2a) to (2c) that (mod 7):

(4a) $$j \equiv W + 2 + [W/4] \equiv G + 2 + [G/4]$$

(4b) $$j \equiv D - 2 + [D/4]$$

(4c) $$j \equiv J + 3 + [J/4].$$

These rules are also found in several treatises[26].

---

22 BN 160 82ᵃ I, 17—19.

23 Vat 119 71ᵃ—84ᵇ; Vindob 6, 4ᵃ—17ᵇ; etc. One also finds simply "John": Berol 84 8ᵇ II, 26; Vat 119 88ᵃ II, 7; etc.

24 Cf. e. g., Hammerschmidt, Äth., p. 134; Weld Blundell, Roy. Chron., p. 504, n. 2 (Decollatio Johannis).

25 BN Abb 37 2ᵃ I, 33 f.: "the Day of John which is the beginning of the days (of the year) and which is also called Beginning of Thoth". In tables, e. g. Berol 84 13ᵇ—15ᵃ I; Vindob 6 19ᵇ—21ᵃ. For tito = Thoth, cf. the lists of month names in different languages (e. g. Princ 5884 12ᵃ III, 17 f.). Dillmann, BM Catal. p. 44, No. 37, 1b renders śarqatito by "novilunium", in flagrant contradiction to the sequence of numerical values of j.

25a BN 160 22ᵇ I, 19; 24ᵃ II, 9/10. Similar EMML 215 46ᵇ II, 22:

The norm $t = 1$ = Wednesday was certainly not so familiar to the scribes as the "feria" $j = 1$ = Sunday. Consequently it is not surprising to find the two parameters confused. We have, e. g., a table for seven 19-year cycles in which the values of t are listed as Day of John[27]. Conversely a lengthy chronological discussion about the life of Christ interprets consistently the values of j and e as t and e and therefore obtains completely incorrect dates[28].

Appendix. The Ṭentyon as Function of c and k (Era G)

The whole Ethiopic computus as represented in the 532-year tables can be considered by itself as a periodic arithmetical structure without any relationship to calendaric concepts. Hence it must be possible to establish on purely number-theoretical grounds all numerical relations which exist between the different tabulated parameters. In the present context, e. g., it is quite natural to investigate the distribution of the values of t (between 1 and 7) in relation to the cycle number c (from 1 to 19) and k (from 1 to 532).

In fact it was exactly this purely arithmetical analysis of the 532-year tables that led me to an understanding of the simple fundamental rules for the dates of Jewish and Christian feast days within the Alexandrian calendar. Having reached this point it is no longer necessary to present the details of the relevant arithmetical procedures. Nevertheless it may be useful to present in Table II the result of such an investigation because it reveals also the basis of a grouping of data in quadruples spaced (mod 532) in 95-year intervals.

Indeed, Table II shows to all $7 \cdot 19 = 133$ pairs of t and c the four corresponding numbers k of the 532-year cycle (in the era G) which belong to the same pair t and c. For example, if we know[29] that Easter falls on VII 26 (i. e., the earliest date) once for $c = 16$, $t = 3$, we see from Table II that the same date occurs in the four years $k = 54$, 149, 244, and 491.

The Ethiopic treatises show awareness of these intervals of 95 years by calling them "cycle of the great epact"[30]. We can observe the same also in the Latin tradition. Not only do the Easter tables of Cyrillus (from 437 to 531) and of Dionysius (from 532 to 626) cover exactly 95

---

27 Cf. s. v. Calendar, Tables, p. 41.

28 BM Add 24995 32ᵃ I, 12—32ᵇ I, 7; cf. s. v. Jesus, p. 172.

29 Cf. above, s. v. Feasts, p. 149.

30 Berol 84 18ᵇ II, 8—9.

Table II

t	c=1	c=2	c=3	c=4	c=5	c=6	c=7	c=8	c=9	c=10
1	58 153 248 495	2 97 192 439	478 41 136 383	422 517 80 327	366 461 24 271	310 405 500 215	254 349 444 159	198 293 388 103	142 237 332 47	86 181 276 523
2	115 210 305 400	59 154 249 344	3 98 193 288	479 42 137 232	423 518 81 176	367 462 25 120	311 406 501 64	255 350 445 8	199 294 389 484	143 238 333 428
3	20 267 362 457	496 211 306 401	440 155 250 345	384 99 194 289	328 43 138 233	272 519 82 177	216 463 26 121	160 407 502 65	104 351 446 9	48 295 390 485
4	77 172 419 514	21 116 363 458	497 60 307 402	441 4 251 346	385 480 195 290	329 424 139 234	273 368 83 178	217 312 27 122	161 256 503 66	105 200 447 10
5	39 134 229 324	515 78 173 268	459 22 117 212	403 498 61 156	347 442 5 100	291 386 481 44	235 330 425 520	179 274 369 464	123 218 313 408	67 162 257 352
6	191 286 381 476	135 230 325 420	79 174 269 364	23 118 213 308	499 62 157 252	443 6 101 196	387 482 45 140	331 426 521 84	275 370 465 28	219 314 409 504
7	1 96 343 438	477 40 287 382	421 516 231 326	365 460 175 270	309 404 119 214	253 348 63 158	197 292 7 102	141 236 483 46	85 180 427 522	29 124 371 466

t	c=11	c=12	c=13	c=14	c=15	c=16	c=17	c=18	c=19
1	30 125 220 467	506 69 164 411	450 13 108 355	394 489 52 299	338 433 528 243	282 377 472 187	226 321 416 131	170 265 360 75	114 209 304 19
2	87 182 277 372	31 126 221 316	507 70 165 260	451 14 109 204	395 490 53 148	339 434 529 92	283 378 473 36	227 322 417 512	171 266 361 456
3	524 239 334 429	468 183 278 373	412 127 222 317	356 71 166 261	300 15 110 205	244 491 54 149	188 435 530 93	132 379 474 37	76 323 418 513
4	49 144 391 486	525 88 335 430	469 32 279 374	413 508 223 318	357 452 167 262	301 396 111 206	245 340 55 150	189 284 531 94	133 228 475 38
5	11 106 201 296	487 50 145 240	431 526 89 184	375 470 33 128	319 414 509 72	263 358 453 16	207 302 397 492	151 246 341 436	95 190 285 380
6	163 258 353 448	107 202 297 392	51 146 241 336	527 90 185 280	471 34 129 224	415 510 73 168	359 454 17 112	303 398 493 56	247 342 437 532
7	505 68 315 410	449 12 259 354	393 488 203 298	337 432 147 242	281 376 91 186	225 320 35 130	169 264 511 74	113 208 455 18	57 152 399 494

years, but Dionysius explicitly discusses this periodicity in the introduction to his tables[31]. The basis of Ethiopic and Latin knowledge, of course, lies in the common Alexandrian origin of their tables.

---

[31] Krusch [1937] p. 63, 26. Cf. also Ideler, Chron. II, p. 263.

Thoth

First month of the Alexandrian-Coptic calendar, coinciding with Ethiopic Maskaram. The Ethiopic spellings for Thoth are tawt[1] or tito[2]; therefore śarqatito[3] is the same as śarqa maskaram[4], i. e., beginning of Thoth or Maskaram, the weekday of Maskaram 1 (i. e. the ṭentyon).

---

[1] Dillmann, Lex. 570.
[2] Princ. 5884 12ᵃ III, 17/18.
[3] E. g., Berol 84 13ᵇ, table II col. 3; Princ 5884 9ᵃ/10ᵇ, table col. 3.
[4] E. g., CSCO 61, p. 22, 3; p. 26, 29. Cf. also above p. 222.

Towers

See Zodiacal signs (p. 232).

Translations

See Arabic, Coptic, Greek (p. 17, 70, 165).

Week and Weekdays

The week (sanbat) of the Ethiopic calendar follows the Jewish tradition. Saturday is called "first Sabbath", Sunday "Christian Sabbath", and the remaining weekdays are derived from the numerals from two to six. Hellenistic planetary names left no trace in the Ethiopic terminology.

Counting which begins with Sunday = 1 is reflected in the norm of the "Day of John", the weekday of Maskaram 1[1], in contrast to the counting of the "Ṭentyon" that begins from Wednesday = 1[2]. The 364-day year of Enoch has the advantage that it always begins with the same weekday, supposedly a Wednesday according to modern theories[3].

---

[1] Cf. s. v. Ṭentyon, p. 222.
[2] Cf s. v. Ṭentyon, p. 218.
[3] Cf. e. g., Milik, Enoch, p. 61.

The association of the number 7 with the week has provided the basis for many more chronological units, e. g. the "Week of Enoch" of 700 years and its multiples[4]. On the other hand, the division of the 364-day year into 52 weeks suggested also the complementary concept of 7 intervals of 52 days each[5]. There is no limit to these numerological speculations.

One finds in our texts rules for determining the weekday (w) of a day (d) in a month $\mu+1$ of a year of ṭentyon (t)[6]. Each completed month contributes 2 days to the weekday (because $30 \equiv 2 \bmod 7$), the date d adds $d-1$ days since the first day of the month, so that

$$w' \equiv t+2\mu+d-1 \quad \bmod 7$$

is the weekday in the norm of t in which Sunday $=5$. In order to change to the norm in which Sunday $=1$, one has to add 3 to w'. Thus one obtains

$$w \equiv t+2\mu+d+2 \quad \bmod 7 \tag{1}$$

which is the rule of the texts.

A more convenient version of (1) is obtained if one formulates its rule for a day d in month μ (instead of $\mu+1$). Then one obtains

$$w \equiv t+2\mu+d \quad \bmod 7. \tag{2}$$

Also this version is attested in our texts[7].

The term "week" implies also the general sense of "multiple of seven". The 532-year tables, e. g., usually count on the margin the 7 67-year intervals which are contained in the whole cycle. In this case the term subāʿe (abbreviated su) is used. Similarly 4 "weeks" (sanbat, abbreviated san) of 7 19-year cycles also total 532 years. These 133-year "weeks" are named in the order of the four Evangelists "of Matthew", "of Mark", etc.[8]

---

[4] Cf. s. v. Jubilees, p. 172.

[5] Vindob 6 22[b] II, 22 to 23[a] I, 8; Neugebauer [1964] p. 66. In BN Abb 37 4[b] I, 29, these intervals (or their endpoints) are called "openings (reḫwa) of the heavens"—for reasons unknown.

[6] Ṭānā 34 101[b] II, 5 to 14 and Vat 119 88[a] II, 19 to 88[b] I, 17 are close parallels. A shorter and garbled version is found in Princ 5884 12[a] I, 16 to 20 and Ṭānā 53 108[b] I, 15 to 19.

[7] Berol 84 39[b] I, 28—II, 17; BN 160 81[a] II, 18—81[b] II, 7.

[8] EMML 215 83[a]—94[b]; Vindob 6 4[a]—14[b]. Cf. also s. v. Evangelists, p. 127.

Weeks of Evangelists

See Evangelists (p. 127).

Weeks of Ezra and Enoch

See Jubilees (p. 172).

Windows

Sun, moon, and some important stars rise and set through the six "gates" at the horizon. But according to the Book of Enoch (ch. 72, 7) 12 "windows" (masākwe) are located above gate 4 in the east where the sun rises in the first month, i. e., in Spring. Furthermore (ch. 72, 3) there are "many windows to the right and to the left" of the gates[1]. The purpose of these windows is not clear[2]; we are only told that the 12 windows eject flames at proper times.

In an Amharic section of BM 816[3] four windows are mentioned, each of which covers 60 kēkros. But there are also 6 windows in the east and 6 in the west which are set in relation to the 12 tribes of Israel.

Finally, under Arabic influence, the windows (as well as the gates) are interpreted as zodiacal signs. Also the planets are somehow connected with 12 windows[4]. Obviously any knowledge of the original concepts has long been lost.

Quite obscure is a statement[5] at the end of a list of the six pairs of months: "IV V 1. III VI 1. etc., XII I 1." saying: "these are the windows of the stars". Apparently one "window" is assigned to each pair of months.

---

[1] Cf. s. v. Gates, p. 161.

[2] A slightly different version, but equally obscure, is found in Berol 84 27[b] I, 15 to II, 3.

[3] BM 816, 19[b] II, 15.

[4] Vat 119, 57[b] I.

[5] BM 816 5[b] on top of a table of lunar mansions.

Winds

See also Maps; Orientation.

In two sections the Book of Enoch describes the directions of the winds and their influences (ch. 34 to 36 and, in a slightly more elaborate version, ch. 76). There are 12 "gates for the winds" (ḫawāḫwa nafāsāt), 3 in each cardinal direction. The winds coming from the 4 central gates are favorable, bringing peace, justice, health, etc., but the winds from the remaining 8 gates, between the cardinal directions, are destructive, bringing hail, locusts, etc.[1].

This simple geometrical pattern should not be taken as evidence of an exact twelve-division of the horizon since we know nothing about the sizes of the gates of the winds.

It may be mentioned that "gates" for the directions of the winds are also known from classical antiquity, e. g. in Galen[2].

---

[1] Enoch, ch. 34, 3 to 76, 4. The same distribution is also found in the Aramaic version; cf. Milik, Enoch, p. 285. Similarly, Berol 84 34ᵃ I, 20 to 34ᵇ II, 3; BM 816 19ᵃ II, 1 to 19ᵇ I, 15; Jerus 268 fol. 3 (Littmann [1902]).

[2] Galen, Comm. Hippocr. Septim., p. 45e.

World

See Eras (p. 116.).

Year

In antiquity the term "year" has very different meanings. To an Egyptian it meant a fixed interval of 365 days. To a Babylonian it represented a time of either 12 or 13 lunar months, each of which could have either 29 or 30 days (depending, in a very irregular fashion, on the actual visibility of the moon). Similar "years" of greatly variable length dominate the calendars of the Greek cities and of the Jews in Palestine. Finally, Augustus enforced in Rome as well as in Egypt a "year" of either 365 or 366 days (each fourth year), the so-called "julian" year. Combined with "months" of a constant length of 30 days, and 5 or 6 epagomenal days, this year became generally used in Roman Egypt. We call it the "Alexandrian" year.

It is important to keep in mind the ambiguity of the concept "year". For example, the term "19-year cycle" has very different meanings for the Babylonian and for the Alexandrian year. In Babylonia (of hellenistic times) it is a definite rule for the arrangement of the 12- and 13-month years[1]; in the Alexandrian (or Ethiopic) computus it means the cyclic

---

[1] Cf. above s. v. Feasts, p. 138.

repetition of certain civil dates with a period of 19 Alexandrian years (e. g., the civil dates of Passover). The "year" of Enoch[2], clearly imitating the Egyptian year, also has a fixed length but its divisibility by 7 excludes any cyclical variation of weekdays—the obvious purpose of its construction.

The architects of the Christian calendar were not concerned with the accuracy of definitions or agreement with empirical data. They completely ignored the brilliant results of contemporary pagan astronomical science and turned instead to the mystical qualities of the primitive cycles on which their "computus" rested. Central among these is the number seven, hallowed by Genesis and thus deeply embedded in the Jewish and Christian calendars. The "weeks", the "jubilees", and the 7000 years to the Second Coming of Christ appear in countless combinations in all of our treatises. Nevertheless connections with contemporary life are unavoidable and the Alexandrian year remains the framework for the 19-year cycle of the Jewish feasts as well as for the Christian civil calendar[3]. Thus we find recorded, e. g., the number of Sundays (365 250) in 7000 years, or the total duration of the world (2 556 750 days)[4]. Abū Shāker considers the Alexandrian year to be the "solar year", in contrast to the "lunar year" of 354 days[5]. Occasionally later scribes mention accurate parameters from Arabic astronomy (usually in Amharic)[6], but which remain, of course, without any influence on the main text[7].

The generally accepted "lunar year" of 354 days is again distinguished by its arithmetical convenience, the systematic alternation be-

---

[2] Book of Enoch, ch. 74, 75, 82; Book of Jubilees, VI, 32 and 38; also in many treatises, e. g., Berol 84 5ª I, 3—15.

[3] This leads to many inconsistencies, e. g., in the Slavonic Enoch between the Enoch year (e. g. the 2 times 182 "thrones" of the sun in ch. 48 (Bonwetsch [1922] p. 43/44; Vaillant, Secrets, p. 109) and the julian year in ch. 14, 1 (Bonwetsch [1922] p. 13; Vaillant, Secrets, p. 91).

[4] E. g., Berol 84 11ª I and II, 3/4; 40ᵇ III, 14—19; Vindob 6 21ª II, 14—17.

[5] BM Add 16252, e. g. 31ª, 19; 32ª, 1. At the beginning of ch. 10, however, the "solar year" is said to contain 365 days (31ª, 11—13), probably because the underlying chronological schemes operate both with Egyptian and Alexandrian years (cf. Neugebauer, Al-Khwarizmi, p. 82 f.).

[6] E. g. Vat 106 92ᵇ II, 12—14; Vat Borg 1 3ᵇ. Cf. also s. v. Arabic, p. 18.

[7] Reminiscent of the useless references by modern historians to accurate parameters.

tween full and hollow months[7a]. The corresponding mean length of 29½ days for the synodic month is in fact quite adequate for calendaric computations over short intervals[8] while the 19-year cycle in the calendar tables provides a much higher accuracy.

Greek astronomy, beginning with Hipparchus in the second century B. C., initiated an entirely new approach to the concept "year" by distinguishing the sidereal from the tropical year. At the same time it had become obvious that the conventional calendar years, be it Egyptian or later Alexandrian years, had no astronomical significance but were useful to provide convenient units of time measurement. Assimilating, furthermore, Babylonian units of time and angular distances resulted in standards that have survived to the present day as units in spherical astronomy.

At first sight it seems as if this development in Greek astronomy did not touch the Ethiopian treatises, simply because of the absence of any theory of solar, lunar or planetary motion. It is therefore of some interest to note some passages which make sense only in relation to the concept "sidereal year".

When a star[9] appears on the eastern horizon just before sunrise (i. e. still in the twelfth hour of night) its visibility is of short duration only. But the progress of the sun in the ecliptic causes day after day earlier risings and thus longer visibility before sunrise. This process continues until the star is visible all night, i. e., until it rises in the first hour of night. From then on the rising slips into daylight and the star becomes less and less visible after sunset until it comes so near to the sun that it remains invisible. This cycle repeats itself until the distance between star and sun is restored to the situation from which we started. This period, which is characterized by the position of the sun in relation to the fixed stars, is therefore called the "sidereal year".

What we have described here in words is represented in our treatises by simple numerical schemes. The "delay" of the rising of the sun with respect to the rising of a star is assumed to amount to two hours each month[10], hence one day in one year. This process is described in 12 single

---

[7a] Book of Enoch, ch. 78, 15 f.; 79, 4 and passim in later treatises.

[8] Cf. the schematic lunar calendar which starts each year with the "maṭqeʿe" (cf. s. v. Feasts, p. 137).

[9] We assume a position not far from the ecliptic (e. g. rising within the "gates"). Otherwise such qualitative considerations can be very misleading.

[10] Berol 84 33ᵇ I, 12 to 21; 40ᵇ III, 27 to 41ᵃ I, 26.

steps from "month 1" (identified with Miyāzyā [VIII] of the civil calendar) to "month 12" (Magābit, i. e. VII)[11].

The association of the "year" with a sidereal period is a very natural approach. Trouble begins, however, as soon as one tries to correlate the sidereal year with some other calendaric unit, e. g., a civil year of 12 months, in which the concept "month" in some way is associated with the moon. Also schematically defined "seasons" will sooner or later fall out of step with sidereal returns. This is exactly what we find in chapter 74 of the Book of Enoch. In the sections 1 to 9 the risings of sun and moon are described in relation to the "gates", a topic mentioned once again in the concluding section 17. In between, however, we find a totally independent discussion, based on two numbers characteristic of the astronomy of Enoch: the lunar year of 354 days and the "solar" year of 364 days, i. e., of four seasons of 91 days each. Furthermore there is introduced silently what in fact is a sexagesimal division of the (sidereal) year, probably inspired by the hellenistic concept of "time degrees". The units are again called "days", a terminology which greatly contributes to obscuring the text. What is meant, however, is very simple[12]: one lunar year amounts to 354 days, one sidereal year to 360 "days"; hence the difference in 5 years is $5 \cdot 6 = 30$ "days", i. e., one "month".

For the explanation suggested here it is important to note that we have direct evidence for a 360-division in the norm[13]

$$24^{h} = 360 \text{ kēkros}.$$

Here the "kēkros" are the exact equivalent of the Greek time degrees (χρόνοι).

The combination of real days with sexagesimal "days" is, of course, only an empty play with numbers, so dear to the wisdom of Enoch. Nevertheless it is clear that the author was trying to establish, or to motivate, some sort of intercalation cycle. Consequently he proceeds[14] by comparing the lunar year with the calendar year of 364 days. Now three lunar years are 1062 days, and three calendar years 1092 days, thus again achieving the desired excess of one month, this time

---

[11] The hours of the rising of the star are always given two consecutive numbers: "12 and 11", "10 and 9", etc. For the beginning of month 1 = VIII cf. s. v. Seasons, p. 207 (5).

[12] Enoch, ch. 74, 10 f.

[13] Berol 84 33[a] II, 16 to III, 3. Cf. also s. v. kēkros and kentros, p. 177 and s. v. Hours, p. 168.

[14] Enoch, ch. 74, 13 to 16.

based on the correct "epact" of 10 days inherent in the Enoch calendar[15]. What follows is an abortive attempt to explain also the usefulness of an 8-year intercalation cycle, i. e. an "octaeteris". To this end the same operation is carried out for $8 = 3 + 5$ years with the obvious result that the difference between 8 civil years and 8 lunar years is 80 days. Here the text ends. Apparently the author of this intrusion knew of the existence of an octaeteris[16] but he did not realize that it could not be applied to the years of Enoch[17].

[15] Explicit mention of an "epact" of 10 days for the calendar of Enoch is found, e. g., in Berol 84 35a I, 12 to 19. Such statements were bound to be "corrected" by later scribes who knew only an "epact" of 11 days.

[16] This has been seen correctly by Charles, Enoch, pp. 201/202, note 74, 13 to 16.

[17] Cf. s. v. Cycles, p. 88.

Zero

Whenever the number zero occurs in tables it is written as a word "nothing" ('albo). No special symbol (as in Greek and Arabic) seems to be known.

Zodiac, Zodiacal Signs

In Ethiopic texts the zodiacal signs appear exclusively in transliteration of their Arabic names[1]. It was also easy to misinterpret the 12 gates of sunrise and sunset as the 12 zodiacal signs; thus one finds expressions like "the sun rises through the gate Aries (ḥamal), etc.[2]

In another terminology, also derived from the Arabic, the zodiacal signs are called "towers" (māḫfad)[3], corresponding to Greek πῦργοι[4].

[1] E. g., Vindob 6 32b I, 1—5. European missionaries introduced Latin forms, e. g. 'akuāriyu (Aquarius): BM 816 8b II, 14; similarly 8b II, 5; 10a I, 15.

[2] Vat 54 88a, 9—89a, 19. Cf. for this text also s. v. Enoch, p. 113 ch. XLVIII and s. v. Seasons, p. 208. The same terminology is used by Abū Shāker (13th cent.), e. g. in chapter 25 (BM Add 16252, 128b, 19).

[3] E. g., BM 816 7b passim; Berol 84 7b II, 26/27; 22b I, 6/7.

[4] Cramer, Anecd. Gr. I, p. 371, 15 f.; 372, 25; 374, 32; CCAG 8, 3 p. 8, f. 79b; p. 10, f. 105; CCAG 9, 2 p. 50 f. 22b (f. 23a passim).

These towers are counted either on two semicircles from 1 to 6, similar to the counting of the gates[5], or in one sequence from 1 to 12[6].

Modern scholars often acted like the Arabs by relating any number 12 immediately to the zodiac. This attitude obscures an important historical situation: primitive astronomy of the Enoch type is not concerned with the orbital motions of sun and moon but concentrates on the easily visible horizon phenomena: heliacal risings or settings, risings amplitudes, first and last visibilities. The step to a moveable sidereal coordinate system belongs to an entirely different level of astronomical knowledge.

---

[5] BM 809 3ᵃ—4ᵃ; BM 816 16ᵃ—17ᵃ (parallel versions). Contamination with the number 30 (days of the sun's travel in each sign) gives the impression of a numbering from 31 to 36.

[6] BM Add 24990 27ᵇ—28ᵇ.

III. BIBLIOGRAPHICAL ABBREVIATIONS

Abbadie, Cat.: Antoine d'Abbadie, Catalogue raisonnée des manuscrits éthiopiens appartenant a Antoine d'Abbadie. Paris 1859.

Abu Šaker, Comp.: Abušaker, Computus (from BM Add 16252; cf. Dillmann, B. M. Cat. p. 40, No. 36).

ACT *see* Neugebauer ACT.

Allegro [1964]: J. M. Allegro, An Astrological Cryptic Document from Qumran. Journal of Semitic Studies 9 (1964) pp. 291—294.

Ambrosiana *see* Galbiati [1957].
see Grébaut [1933].

Assfalg-Krüger, WB: Julius Assfalg - Paul Krüger, Kleines Wörterbuch des Christlichen Orients. Harrassowitz, Wiesbaden, 1975.

Bar Šinaya, Chron. (Chabot): I. B. Chabot, Eliae Metropolitae Nisibeni, Opus chronologicum. II. CSCO 63** (Scriptores Syri 24 = ser. III t. 8 Vessio) Louvain 1910 (repr. 1954).

Bar Šinaya, Chron. (Delaporte): L.-J. Delaporte, Chronographie d'Élie Bar Šinaya, Métropolitain de Nisibe. Bibl. de l'École des Hautes Études 181, Paris 1910.

Barthélemy [1952]: D. Barthélemy, O. P., Notes en marge de publications récentes sur les manuscrits de Qumran. Revue Biblique 59 (1952) pp. 187 to 218.

Basset, Apocr. X: René Basset, Les Apocryphes Éthiopiens. X. La sagesse de Sibylle. Paris 1900.

Berg, Karte: Wilhelm Berg, Historische Karte des alten Ägypten. Verlag Hans Richarz, 1973.

Berlin *see* Dillmann, Berlin HSS.

Bernhard, Chron.: Ludger Bernhard, Die Chronologie der Syrer. Österr. Akad. d. Wissenschaften, Philos. hist. Kl. Sitzungsberichte 264 (1969) Abh. 3.

Bibliothèque Nationale *see* Zotenberg, B. N.
Collection d'Abbadie *see* Abbadie, Cat.
Collection Griaule *see* Cat. Griaule.

Bīrūnī, Chron.: The Chronology of Ancient Nations. trsl. Edward Sachau, London 1879.

Bodleian *see* Dillmann, Bodl. Cat.
see Ullendorff, Bodl. Cat.

Bonwetsch, G. N.
[1896]: Das slavische Henochbuch. Abhandlungen der königl. Gesellschaft d. Wissensch. zu Göttingen, Philol.-histor. Kl., N. F. 1, Nr. 3 (Berlin, Weidmann, 1896).
[1922]: Die Bücher der Geheimnisse Henochs. Das sogenannte slavische Henochbuch. TU 44, 2 (1922).

British Museum *see* Dillmann, Br. Mus. Cat.
see Wright, Br. Mus. Cat.

Burmester [1951]: O. H. E. Hadji Burmester, Rites and Ceremonies in the Coptic Church. The Eastern Churches Quarterly 9 (1951/52) pp. 306—319.

Cat. Griaule
Catalogue des Manuscrits Éthiopiens de la Collection Griaule
I. Bibl. Nat., Dep. des manuscr., Paris 1938;
II. Miscellanea africana Lebaudy 3, Paris 1941;
III. Travaux et Mém. de l'Inst. d. Ethnol. 30, Paris 1944;
IV. Bibl. Nat., Dep. des manuscr., Paris 1954
(I, II, III ed. Grébaut; IV, ed. Strelcyn).

CCAG: Catalogus Codicum Astrologorum Graecorum.

Cerulli, Enrico
Et. in Pal.: Etiopi in Palestina, Storia della comunità etiopica di Gerusalemme. Roma 1943—1947, 2 vols.
Lett.: Storia della letteratura etiopica. 3rd ed., Milano 1968.
[1965]: Manoscritti etiopici della Chester Beatty Library in Dublino. Atti della Accademia Nazionale dei Lincei, Anno 362 (1965) Serie 8, Memorie. Classe di Scienze morali, storiche e filologiche 11 (1962 to 1965) pp. 277—323.
[1973]: Tiberius and Pontius Pilate in Ethiopian Tradition and Poetry. Proceedings of the British Academy 59 (1973) pp. 141—158.

Chabot *see* Bar Šinaya.

Chaine, Marius
Abbad.: Bibliothèque Nationale, Catalogue des Manuscrits Éthiopiens de la collection Antoine d'Abbadie. Paris 1912.
Chron.: La chronologie des temps chrétiens de l'Égypte et de l'Éthiopie. Paris, Geuthner, 1925.
Gramm.: Grammaire éthiopienne. Nouvelle édition. Beyrouth, Imprim. Cathol., 1938.
Mondon-V.: Catalogue des manuscrits éthiopiens de la collection Mondon-Vidailhet (Bibliothèque Nationale, Départ. des Manuscrits) Paris 1913.
[1912]: Inventaire sommaire des manuscrits éthiopiens de Berlin acquis depuis 1878. ROC 17 (1912), pp. 45—68.
[1931]: Le Chronicon Orientale de Butros ibn ar-Rahib et l'histoire de Girgis el-Makim. ROC 28 (1931/32) pp. 390—405.
[1936]: Les manuscrits éthiopiens de la Bibliothèque des RR. PP. Capucins a Toulouse. Aethiopica 5 (1936), pp. 8—16, pp. 23—29.

Charles, Robert Henry
Chronicle: The Chronicle of John, Bishop of Nikiu. Text and Translation Society, London 1916.
Enoch: The Book of Enoch, trsl. from Professor Dillmann's Ethiopic Text. Oxford 1893.
Ethiop. Jubil.: The Ethiopic Version of the Hebrew Book of Jubilees. Anecdota Oxoniensia, Oxford 1895.
Jubilees: The Book of Jubilees or The Little Genesis, trsl. from the editor's Ethiopic text. Oxford 1902.
[1921]: The Date and Place of Writing of the Slavonic Enoch. J. of Theological Studies 22 (1921) pp. 161—163.

Charles-Morfill, Enoch: R. H. Charles - W. R. Morfill, The Book of the Secrets of Enoch, trsl. from the Slavonic. Oxford 1896.

Conti Rossini, Carlo

L. Ax.: Liber Axumae. CSCO 54, 58 (1909; reprinted 1954).

[1899, 1]: Note per la storia letteraria Abissina. Rendiconti della Reale Accademia dei Lincei, Classe di scienze morali, storiche e filologiche. Ser. 5, vol. 8 (1899) pp. 197—220; pp. 263—285.

[1899, 2]: Manoscritti ed opere abissine in Europa. Rendiconti della Reale Accademia dei Lincei, Classe di scienze morali, storiche e filologiche. Ser. 5, vol. 8 (1899), pp. 606—637.

[1900]: Il Gadla Filpos e il Gadla Johannes di Dabra Bizan. Atti della R. Accademia dei Lincei, Anno 297 (1900), Serie 5ª. Classe di scienze morali, storiche e filologiche, vol. 8. Parte 1ª: Memorie; p. 62—170.

[1904]: I manoscritti etiopici della missione cattolica di Cheren. Rendiconti della Reale Accademia dei Lincei, Classe di scienze morali, storiche e filologiche, ser. 5, vol. 13 (1904), pp. 233—255, pp. 261—286.

[1912—1915]: Notice sur les manuscrits éthiopiens de la Collection d'Abbadie. J. As. sér. 10, vol. 19 (1912) pp. 551—578; vol. 20 (1912) pp. 5 to 72; sér. 11, vol. 2 (1913) pp. 5—64; vol. 6 (1915) pp. 189—238, pp. 447—493.

[1940]: Il libro della rivelazione dei misteri attribuito a Tolomeo. RSO 19 (1940) p. 65—80.

Convegno 1959: Atti del Convegno Internazionale di Studi Etiopici. Roma 2—4 Aprile 1959. Accademia Nazionale dei Lincei, Anno 357 (1960) No. 48. Problemi attuali di Scienze e di Cultura.

Cosmas, Top. Chrét.: Cosmas Indicopleustès, Topographie chrétienne, ed. Wanda Wolska-Conus. Sources Chrétiennes 141, 159, 197 (1968, 1970, 1973).

CRAIBL: Académie des Inscriptions et Belles-Lettres, Comptes Rendus des Séances. Paris.

CSCO: Corpus Scriptorum Christianorum Orientalium

My references give the volume numbers of the reprint edition. The following concordance should make it possible to use also the original numbering of the "Scriptores Aethiopici":

CSCO	Script.	CSCO	Script.
20/21	3/4	54/58	24/27
22/23	5/6	61/66	28/29
24/25	7/8	68	30
26/27	9/10	138/139	31/32
30/31	13/14	221/222	41/42
33/34	16/17	235/236	43/44

da Leonessa *see* Mauro da Leonessa

Delaporte *see* Bar Šinaya

Dexinger, Apok.: Ferdinand Dexinger, Henochs Zehnwochenapokalypse und offene Probleme der Apokalypsenforschung. Leiden, Brill, 1977 (Studia Post-Biblica 29).

Didascalia: J. M. Harden, The Ethiopic Didascalia. Transl. of Christ. Lit Ser. 4, Oriental Texts. London 1920.
Dillmann, August
Berlin HSS: Die Handschriften-Verzeichnisse der Königlichen Bibliothe zu Berlin. Bd. 3. Verzeichniss der abessinischen Handschriften. Berli 1878.
Bodl. Cat.: Catalogus codicum manuscriptorum Bibliothecae Bodleiana Oxoniensis, Pars VII. Oxford 1848.
Br. Mus. Cat.: Catalogus Codicum Manuscriptorum Orientalium qui i Museo Britannico asservantur. Pars III. Codices Aethiopicos amplec tens. London 1847.
Gramm.: Grammatik der äthiopischen Sprache. 2te Aufl., Leipzig 189 [Neudruck, Graz 1959].
Henoch: Liber Henoch, Aethiopice, ad quinque codicum fidem editus cum variis lectionibus. Leipzig 1851.
Lex.: Lexicon linguae aethiopicae. Leipzig 1865 (Reprint: New Yor 1955).
Doresse [1962]: Jean Doresse, Littérature éthiopienne et littérature occiden tale au moyen-âge. Bulletin de la Société d'Archéologie Copte 16 (1961 t 1962) pp. 139—159.
Dornseiff, AMM: Franz Dornseiff, Das Alphabet in Mystik und Magie Stoicheia 7 (1922).
Duchesne [1880]: L. Duchesne, La question de la paque au Concile de Nicée Revue des Questions Historiques 28 (1880) pp. 5—42.
Dupont-Sommer [1977]: André Dupont-Sommer, Trente années de recher ches sur les manuscrits de la mer Morte (1947—1977). CRAIBL 197 p. 659—677.

Elias of Nisibis *see* Bar Šinaya.
Enoch: *see* Charles
see Flemming
Ephraim Isaac [1976]: Catalogue of the Ethiopic (Ge'ez) MSS in the Man uscript Library of the Armenian Patriarchate of Jerusalem. Muséon 8 (1976) pp. 179—194.
Eusebius, H. E.: Eusèbe de Césarée, Histoire Ecclésiastique, ed. G. Bardy Sources Chrétiennes 31, 41, 55, 73 (Paris 1952—1960).
Ewald [1844]: H. Ewald, Über die äthiopischen Handschriften zu Tübingen Zeitschr. für die Kunde des Morgenlandes 5 (1844) pp. 164—201.

Flemming, Johannes
Henoch: Das Buch Henoch. Äthiopischer Text. TU 22, 1 (1902).
[1906]: Die neue Sammlung abessinischer Handschriften auf der König lichen Bibliothek zu Berlin. Zentralblatt für Bibliothekswesen 2 (1906) p. 7—21.
Flemming-Radermacher, Henoch: Joh. Flemming - L. Radermacher, Da Buch Henoch. GCS 5, Leipzig, Hinrichs, 1901.
Floëri-Nautin, Hom. Pasc.: F. Floëri - P. Nautin, Homélies Pascales III Une homélie Anatolienne sur la date de Paques en l'an 387. Source Chrétiennes 48 (1957).

Fotheringham [1922]: J. K. Fotheringham, The Easter Calendar and the Slavonic Enoch. J. of Theological Studies 23 (1922) p. 49—56.

Fumagalli, BE: Giuseppe Fumagalli, Bibliografia etiopica. Milano 1893 (cf. Zanutto BE I p. 26 No. 45).

Galbiati [1957]: Enrico Galbiati, I manoscritti etiopici dell'Ambrosiana (breve inventario). Studi in Onore di Carlo Castiglioni, Milano, 1957, p. 337—353.

GCS: Die Griechischen Christlichen Schriftsteller der ersten drei Jahrhunderte.

Georgy Sobhy Bey [1942]: The Coptic calenderical computation and the system of epacts . . . ascribed to Abba Demetrius the XIIth Patriarch of Alexandria. Bulletin de la Société d'Archéologie Copte 8 (1942) p. 169 to 199 (Arabic text: vol. 9, p. 237—252).

Ginzel, Hdb.: F. K. Ginzel, Handbuch der mathematischen und technischen Chronologie. 3 vols., Leipzig, Hinrichs, 1906, 1911, 1914.

Goldschmidt, Ab. Hs.: Lazarus Goldschmidt, Die Abessinischen Handschriften der Stadtbibliothek zu Frankfurt am Main. (Rüppellsche Sammlung) Berlin, 1897.

Goldstine, Tables: Herman H. Goldstine, New and Full Moons, 1001 B. C. to A. D. 1651. Amer. Philos. Soc., Memoirs 94 (1973).

Graf, CAL: Georg Graf, Geschichte der christlichen arabischen Literatur. II. Die Schriftsteller bis zur Mitte des 15. Jahrh. Studi e Testi 133, Città del Vaticano, 1947.

Grébaut, Silvain

[1913]: Les sept cieux et les sept cercles de la terre. ROC 18 (1913) p. 204—206.

[1917]: Les manuscrits éthiopiens. ROC 20 (1915—1917) p. 408—415.

[1918, 1]: Table de Comput et de Chronologie. ROC 21 (1918—1919) p. 323—330.

[1918, 2]: Table des levers de la lune pour chaque mois de l'année. ROC 21 (1918—1919) p. 422—428.

[1918, 3]: Variations de la durée des jours et des nuits pour chaque mois de l'année. ROC 21 (1918—1919) pp. 429—432.

[1920]: Calculs et tables relatifs au comput. ROC 22 (1920—1921) pp. 212—220.

[1933]: Catalogue des manuscrits éthiopiens de la Bibliothèque Ambrosienne. ROC 29 (1933—1934) pp. 3—32.

Grébaut-Tisserant, Cod. Aeth. Vat. et Borg.: Silvain Grébaut - Eugene Tisserant, Codices Aethiopici Vaticani et Borgiani, Bibl. Vaticana, I, 1935, II, 1936.

Griaule [1934]: Marcel Griaule, Notes sur l'arithmomancie éthiopienne. Journal de la Société des Africanistes 4 (1934) pp. 25—31.

Griaule, Collection *see* Cat. Griaule.

Grumel, V.

Chron.: La Chronologie. Taité d'Études Byzantines I, Paris 1958.

[1960]: Le problème de la date pascale aux IIIe et IVe siècles. Revue des Études Byzantines 18 (1960) pp. 161—178.

Guidi, Ignazio
Voc.: Vocabulario Amarico-Italiano. Pubblicazioni dell'Istituto per l'Oriente, Roma, 1953.
[1896]: Il "Marḥa 'Ewûr". Rendiconti della Reale Accademia dei Lincei, Classe di scienze morali, storiche e filologiche. Ser. 5, vol. 5 (1896) pp. 363—385.

Hammerschmidt, Ernst
Äth.: Äthiopien, Christliches Reich zwischen Gestern und Morgen. Wiesbaden, Harrassowitz, 1967.
Kal.: Äthiopische Kalendertafeln. Wiesbaden, Steiner, 1977.
Ṭānā: Äthiopische Handschriften vom Ṭānāsee.
1: Reisebericht und Beschreibung der Handschriften in dem Kloster des Heiligen Gabriel auf der Insel Kebrān.
2: Die Handschriften von Dabra Māryām und von Rēmā.
Verzeichnis der Orientalischen Handschriften in Deutschland 20, 1 und 20, 2. Wiesbaden, Steiner, 1973 und 1977.
Harden, Didasc.: J. H. Harden, The Ethiopic Didascalia. Translations of Christian Literature, Series IV. Oriental Texts, 1920.
Hennecke-Schneemelcher, NT Apocr.: Edgar Hennecke - Wilhelm Schneemelcher, New Testament Apocrypha. English transl. by R. McL. Wilson. 2 vols., Westminster Press, Philadelphia, 1963.
Henoch: *see* Dillmann.
Huntingford, Charters: G. W. B. Huntingford, The Land Charters of Northern Ethiopia. Addis Ababa - Oxford Univ. Press, 1965.

Isaac: *see* Ephraim Isaac.

J. As.: Journal Asiatique.
Jerus. Arm.: Manuscript Library of the Armenian Patriarchate of Jerusalem (cf. Ephraim Isaac [1976]).
Jones, Beda: Charles W. Jones, Bedae Opera de Temporibus. Mediaeval Acad. of America, Cambridge, Mass. 1943.

Kamil [1957]: Murad Kamil, Les manuscrits éthiopiens du Sinaï. Annales d'Éthiopie 2 (1957) pp. 83—90.
Kennedy, Survey: E. S. Kennedy, A Survey of Islamic Astronomical Tables. Trans. Amer. Philos. Soc., N. S. 46, 2 (1956).

Leonessa: *see* Mauro da Leonessa.
Liber Axumae: *see* Conti Rossini, L. Ax.
Lifchitz [1940] = TMR: Déborah Lifchitz, Textes éthiopiens magico-religieux. Université de Paris. Travaux et Mémoires de l'Institut d'Ethnologie 38 (1940).
Littmann, Enno
[1902]: Aus den abessinischen Klöstern in Jerusalem. ZA 16 (1902) pp. 102—124, pp. 363—388.
[1908]: Abessinische Glossen. ZA 21 (1908) pp. 50—109.

Löfgren, Oscar
Kat.: Katalog über die äthiopischen Handschriften in der Universitätsbibliothek Uppsala. Stockholm, Almquist and Wiksell 1974 (= Acta Bibliothecae R. Universitatis Upsaliensis 18).
[1972]: Der Spiegel des Salomo, ein äthiopischer Zaubertext. Festschrift Widengren. Supplement to Numen 21, I (1972), pp. 208—223.
Ludolf, Comm.: Iobi Ludolfi alias Leutholf dicti ad suam Historiam Aethiopicam antehac editam Commentarius. Frankfurt a. M. 1691.
Lundmark [1921]: Knut Lundmark, Suspected New Stars recorded in old chronicles and among recent meridian observations. Publications of the Astronomical Society of the Pacific 33 (1921), pp. 225—238.

Macomber, William F.
Cat.: A Catalogue of Ethiopian Manuscripts microfilmed for the Ethiopian Manuscript Microfilm Library, Addis Ababa, and for the Hill Monastic Manuscript Library, Collegeville. St. John's Abbey and University, Collegeville, Minn., 1975, 1976, 1978.
[1975]: A report on the Ethiopian Manuscript Microfilm Library. Le Muséon 88 (1975), pp. 397—403.
Maimonides, Code: The Code of Maimonides. Book III, Treatise 8. Sanctication of the New Moon. Transl. Solomon Gandz, Suppl. J. Obermann, Comm. O. Neugebauer. Yale Univ. Press, 1956 (Yale Judaica Series 11).
Marḥa 'Ewûr: *see* Guidi [1896].
Maunder [1918]: A. S. D. Maunder, The Date and Place of Writing of the Slavonic Book of Enoch. The Observatory 41 (1918) pp. 309—316.
Mauro da Leonessa
Diss. Cron.: Dissertazioni Cronologiche. Cronologia e calendario etiopico. La tavola pasquale di Anatolio. Tivoli 1934.
[1943]: Un trattato sui calendario redatto al tempo di re 'Amda-Ṣyon I. Rassegna di Studi Etiopici (1943) pp. 302—326.
Milik, Enoch: J. T. Milik, The Books of Enoch. Aramaic Fragments of Qumrân Cave 4. Oxford, Clarendon Press, 1976.
Molsdorf, Chr. Symb.: Wilhelm Molsdorf, Christliche Symbolik der mittelalterlichen Kunst. 2te Aufl., Leipzig 1926 (reprinted Graz 1968).
Monastic Manuscript Microfilm Library: *see* Macomber.
Morgenstern [1955]: Julius Morgenstern, The Calendar of the Book of Jubilees, its Origin and its Character. Vetus Testamentum 5 (1955) pp. 34—76.

Neugebauer, O.
ACT: Astronomical Cuneiform Texts, 3 vols., London, Lund Humphries, 1955.
Al-Khwār: The Astronomical Tables of Al-Khwārizmī. Danske Videnskabernes Selskab, Hist.-filos. Skrifter 4, Nr. 2 (1962).
HAMA: A History of Ancient Mathematical Astronomy. 3 vols., Springer, Heidelberg-New York, 1975.
[1959]: Regula Philippi Arrhidaei. Isis 50 (1959) p. 477 f.

[1962]: Über griechische Wetterzeichen und Schattentafeln. Österr. Akad. d. Wiss., Philos.-Hist. Kl., SB 240, 2 (1962) p. 27—44.
[1964]: Notes on Ethiopic Astronomy. Orientalia 33 (1964) p. 49—71.
[1975]: A Greek World Map. Hommages à Claire Préaux. Bruxelles 1975, pp. 312—317.
[1976]: Ṭentyon. Orientalia 44 (1976) p. 487 f.

Oxf. Dict.: The Oxford Dictionary of the Christian Church. Oxford University Press, 2nd ed. revised 1974.

Patr. Or.: Patrologia Orientalis.
Pauly-Wissowa, RE: Realencyclopädie der classischen Altertumswissenschaft.
Pearson, OM: J. D. Pearson, Oriental Manuscripts in Europe and North America. Bibliotheca Asiatica 7, Inter Documentation Company, Zug. 1971.
Perruchon, Zar'a Yā'eqōb: Jules Perruchon, Les chroniques de Zar'a Yā'eqōb et de Ba'eda Māryām, rois d'Éthiopie de 1434 a 1478. Bibliothèque de l'École Pratique des Hautes Études, Fasc. 93 (1893).
Piper, Kalend.: Ferdinand Piper, Karls des Großen Kalendarium und Ostertafel nebst einer Abhandlung über die lateinischen und griechischen Ostercyklen des Mittelalters. Berlin 1858.
Polotsky [1964]: H. J. Polotsky, Aramaic, Syriac, and Ge'ez. J. of Semitic Studies 9 (1964) pp. 1—10.

Rassam, Theodore: Hormuzd Rassam, Narrative of the British Mission to Theodore, King of Abyssinia. London 1869, 2 vols.
Renaudot, Hist. Patr.: Eusèbe Renaudot, Historia Patriarcharum Alexandrinorum Jacobitarum ab Marco usque ad finem saeculi XII . . ., Paris 1713.
Rhodokanakis [1906]: N. Rhodokanakis, Die äthiopischen Handschriften der k. k. Hofbibliothek zu Wien. Sitzungsberichte der kais. Akademie der Wissenschaften in Wien, Philos.-hist. Kl. 151, 4.
Richard, Marcel
[1966]: Notes sur le comput de cent-douze ans. Mélanges Venance Grumel, Revue des Études Byzantines 24 (1966) pp. 257—277.
[1974]: Le comput pascal par octaétéris. Le Muséon 87 (1974) pp. 307 to 339.
Riessler, Altjüd. Schrifttum: Paul Riessler, Altjüdisches Schrifttum außerhalb der Bibel. Wissenschaftliche Buchgesellschaft, Darmstadt 1966.
ROC: Révue de l'Orient Chrétien.
RSE: Rassegna di Studi Etiopici.
RSO: Rivista degli Studi Orientali.
Rühl, Chron.: Franz Rühl, Chronologie des Mittelalters und der Neuzeit. Berlin 1897.
Rüppell, Reise: Eduard Rüppell, Reise in Abyssinien. 2 vols., Frankfurt a. M., 1838/1840.

Sachs, A.

[1952]: Babylonian Horoscopes. J. of Cuneiform Studies 6 (1952) pp. 49 to 75.

[1976]: The Latest Datable Cuneiform Tablets. Kramer Anniversary Volume 1976, pp. 379—398.

Saliba [1970]: George A. Saliba, Easter Computation in Medieval Astronomical Handbooks. Al-Abhath 23 (1970) pp. 179—212 [Amer. Univ. Beirut].

Scaliger: Joseph Scaliger, De emendatione temporum. 1583, 1598, 1629 (I am quoting from the [posthumous] edition of 1629).

Schram, Tafeln: Robert Schram, Kalendariographische und Chronologische Tafeln. Leipzig, Hinrichs, 1908.

Schroeter, Kanon: J. Fr. Schroeter, Spezieller Kanon der zentralen Sonnen- und Mondfinsternisse, welche innerhalb des Zeitraums von 600 bis 1800 n. Chr. in Europa sichtbar waren. Kristiania 1923.

Schwartz, E.

Ostert.: Christliche und jüdische Ostertafeln. Abhandlungen d. Kgl. Gesellschaft d. Wissensch. zu Göttingen. Philol.-Histor., Kl., N.F. 8 Nr. 6 (1905).

[1904/8]: Zur Geschichte des Athanasius. Nachrichten von der Kgl. Gesellschaft d. Wiss. zu Göttingen, Philol.-Hist. Kl. I—III: 1904, pp. 333—401; IV: 1904, pp. 518—547; V: 1905, pp. 164—187; VI: 1905, pp. 257—299; VII: 1908, pp. 305—374. Reprinted in Gesammelte Schriften Bd. 3 (Berlin 1959) pp. 1—187.

[1906]: Osterbetrachtungen. Zeitschrift für die neutestamentliche Wissenschaft 7 (1906) pp. 1—33. Reprinted in Gesammelte Schriften Bd. 5 (Berlin 1963) pp. 1—41.

Ševčenko [1967]: Ihor Ševčenko, The early period of the Sinai Monastery in the light of its inscriptions. Dumbarton Oaks Papers 20 (1967) pp. 255 to 264.

Sobhy *see* Georgy Sobhy Bey.

Strelcyn, Stefan

B. M.: Catalogue of Ethiopian Manuscripts in the British Library, acquired since the year 1877, London 1978.

Cat. John Rylands: Catalogue of Ethiopic Manuscripts in the John Rylands University Library in Manchester, Manchester Univ. Press, 1974.

C. C. R.: Catalogue des manuscits éthiopiens de l'Accademia Nazionale dei Lincei, Fonds Conti Rossini et Fonds Caetani ..., Accademia Nazionale dei Lincei, Roma 1976.

[1955]: Priéres magiques éthiopiennes pour délier les charmes. Polska Akademia Nauk, Rocznik Orientalistyczny 18, Warszawa 1955.

Strobel, Osterkal.: August Strobel, Ursprung und Geschichte des frühchristlichen Osterkalenders. TU 121. Akademie-Verlag, Berlin 1977.

Ṭānā: *see* Hammerschmidt, Ṭānā.

Toulouse *see* Chaine [1936].

TU: Texte und Untersuchungen zur Geschichte der altchristlichen Literatur.

Ullendorff, Edward
Bodl. Cat.: Catalogue of Ethiopian Manuscripts in the Bodleian Library. Vol. II, Oxford 1951.
Ethiop.: The Ethiopians, an Introduction to Country and People. London, Oxford, Univ. Press, 1960.

Ullendorff-Wright, Cat. Cambridge: Edward Ullendorff - Stephan G. Wright, Catalogue of Ethiopian Manuscripts in the Cambridge University Library, Cambridge University Press, 1961.

Vaillant, Secrets d'Hénoch: A. Vaillant, Le Livre des Secrets d'Hénoch. Texte slave et traduction française. Textes publiés par l'Institut d'Études slaves 4 (1952).

van de Vyver [1957]: A. van de Vyver, L'évolution du Comput Alexandrin et Romain du 3e au 5e Siècle. Revue d'Histoire Ecclésiastique 52 (1957) pp. 5—25.

Vatican, Catal.: *see* Grébaut-Tisserant, Cod. Aeth. Vat. et Borg.

Velat, Meʿeraf: Bernard Velat, Études sur le Meʿerāf commun de l'office divin éthiopien. Patr. Or. 33 (1966).

Viraud [1967]: Gérard Viraud, La procession des deux fêtes de la Croix et du Dimanche des Ramaux dans l'église Copte d'après un manuscrit du Monastère al-Baramûs et du Monastère al-Muḥarraq. Bulletin de la Société d'Archéologie Copte t. 19 (1967/68) pp. 211—226.

Weld Blundell, Roy. Chron.: H. Weld Blundell, The Royal Chronicle of Abyssinia 1769—1840, with Translation and Notes. Cambridge Univ. Press, 1922.

Wright, William
Br. Mus. Cat.: Catalogue of the Ethiopic Manuscripts in the British Museum acquired since the year 1847. London 1877.
[1870]: List of the Magdala Collection of Ethiopic manuscripts in the British Museum. ZDMG 24 (1870) pp. 599—616.

ZA: Zeitschrift für Assyriologie und verwandte Gebiete.

Zanutto, BE: Silvio Zanutto, Bibliografia etiopica in continuazione alla "Bibliografia etiopica" di G. Fumagalli. (2 parts): Roma 1936, 1932.

ZDMG: Zeitschrift der Deutschen morgenländischen Gesellschaft.

Zotenberg, H.
[1883]: Chronique de Jean, évéque de Nikiou. Notices et Extraits des Manuscrits de la Bibliothèque Nationale, 24 (1883), pp. 125—605[1].
B. N.: Catalogue des manuscrits éthiopiens (Gheez et Amharique) de la Bibliothèque Nationale, Paris 1877.

---

[1] Also extant as an independent publication (Paris, Imprimerie National, 1883, with no reference to the Notices et Extraits) with page numbers 120 lower than in the Notices et Extraits. My references are always to the Notices et Extraits.

IV,1. TEXTS

Almost all the texts utilized for the present study are unpublished. The following list is intended to tell the reader where he can find at least a general description of the texts in question.

Abbad 37	Abbadie, Cat. p. 45; Chaine, Abbad. p. 26; Conti-Rossini [1912—15] No. 121 ([1913] p. 5).
Ambros B 22	Galbiati [1957] p. 342 No. 4
Berol 84	Dillmann, Berlin MSS p. 73 (or. qu. 370)
Berol 991	(or. oct.) Chaine [1912] p. 60, No. 56
Berol 994	(or. qu.) Chaine [1912] p. 52, No. 25; Flemming [1906] p. 12, No. 15
Berol 1008	(or. qu.) Chaine [1912] p. 55, No. 39
Berol 1009	(or. qu.) Chaine [1912] p. 56, No. 40; Flemming [1906] p. 15, No. 39, 5
BM 567	Wright, Cat. p. 117 No. 180
BM 570	Wright, Cat. p. 118 No. 183
BM 734	Wright, Cat. p. 203 No. 312
BM 754	Wright, Cat. p. 221 No. 331
BM 809	Wright, Cat. p. 290 No. 383
BM 815	Wright, Cat. p. 226 No. 339[1]
BM 816	Wright, Cat. p. 321 No. 397
BM 827	Wright, Cat. p. 297 No. 390
BM Add 16188	Dillmann, Cat. p. 4 No. 5
BM Add 16220	Dillmann, Cat. p. 10 No. 11
BM Add 16252	Dillmann, Cat. p. 40 No. 36
BM Add 19658	Wright, Cat. p. 110 No. 163
BM Add 24990	Wright, Cat. p. 11 No. 13
BM Add 24995	Wright, Cat. p. 322 No. 398
BN Nos. 1 to 170	Bibliothèque Nationale, numbers as in Zotenberg, Cat.
BN Abb	Bibliothèque Nationale, numbers as in Abbadie, Cat.
Bodl 16	Dillmann, Bodl. Cat. p. 29
Bodl 21	Dillmann, Bodl. Cat. p. 36; Ullendorff, Bodl. Cat. p. 29
Bodl 36	Ullendorff, Bodl. Cat. p. 1
Borg	numbers as in Vat. Cat.
Cambr Add 1570	Ullendorff-Wright, Cat. p. 3 No. 2
Cheren	Conti-Rossini [1904]
EMML	Ethiopian Manuscript Microfilm Library; numbers as in Macomber, Cat.
Frankfurt	numbers as in Goldschmidt, Ab. Hs.
Griaule	numbers as in Cat. Griaule
Jerus Arm.	cf. Ephraim Isaac [1976]
Jerus Littm.	cf. Littmann [1902]

[1] Partial translation in Weld Blundell, Roy. Chron. p. 496—505.

Marḥa 'E	cf. Guidi [1896]
Princ 5884	Princeton University Library (catal. by Ephraim Isaac in preparation)
Scaliger	cf. Scaliger, De emendatione temporum (1629); checked with the originals.
Stockh 29b	Löfgren, Kat. p. 155
Ṭānā	numbers as in Hammerschmidt, Ṭānā
Toulouse	numbers as in Chaine [1936]
Upps 3	Löfgren, Kat. p. 29 No. 3
Upps 42	Löfgren, Kat. p. 131 No. 39
Vat	same numbers as in Grébaut-Tisserant, Cod. Aeth. Vat.
Vat 61	Mauro da Leonessa [1943]
Vindob 6	Rhodokanakis [1906] p. 81 No. 25
Vindob 16	Rhodokanakis [1906] p. 3 No. 1

IV,2. PASSAGES CITED

m, n means page m, note n

V. SUBJECT INDEX

m, n means page m, note n

VI. NUMERICAL PARAMETERS

1. Decimal

1064	$= 2 \cdot 532$
1096	$= 3 \cdot 365 + 1$
133	$= 532/4$
143	$143 \cdot 49 = 7007$, 143 Jubilees $\sim$ cycle of Ezra
1463	$= 7 \cdot 11 \cdot 19$
150	number of Psalms
153	$153 \equiv 1 \bmod 19$; G $153 =$ J 505, $505 \equiv 1 \bmod 28$; $\equiv 1$ for epact and solar cycle
1596	$= 3 \cdot 532$
194180	$532 \cdot 365 = 194180^{d}$
194308	$532^{y} = 194308^{d}\ 22;9^{h}$ (Borg 1 3^{b}, 1—15)
194311	$28 \cdot 235^{m} = 194311^{d}\ 7;9,6,20^{h}$ (Borg 1 3^{b}, 1—15)
209	$= 11 \cdot 19$
2128	$= 4 \cdot 532$
22	number of books in the AT
228	Arabic lunar cycle, $= 12 \cdot 19$
235	19-year cycle, $= 12 \cdot 19 + 7$
247	$= 13 \cdot 19 = 532 - 3 \cdot 95$
2548000	$= 364 \cdot 7000$
2556750	$= 365.25 \cdot 7000$
2660	$= 5 \cdot 532$
275	duration of pregnancy
28	solar cycle
318	number of Fathers at the Council of Nicaea
3192	$= 6 \cdot 532$
343	$= 7 \cdot 49$, 7 Jubilees, Week of Jeremy
361	$= 19 \cdot 19$
364	$= 4 \cdot 91 = 13 \cdot 28 = 52 \cdot 7$, year of Enoch
365250	number of Sundays in 7000 years
368	$368 \cdot 19 = 7000 - 8$
3724	$= 7 \cdot 532$
378	number of verses of Psalms for the epagomenal days
4256	$= 8 \cdot 532$
4300	from Enoch to Birth of Christ
446	Week of Jeremy (?)
4788	$= 9 \cdot 532$
48	in 532 years Easter falls 48 times in VII
483	$= 7 \cdot 60$, 7 Weeks of Daniel
484	in 532 years Easter falls 484 times in VIII
49	Jubilee
532	$= 4 \cdot 7 \cdot 19$, great lunar cycle
5418	number of verses in 150 Psalmes

5852	= 11 · 532
6384	= 12 · 532
69	Week of Daniel
6916	= 13 · 532 = 19 · 364
6935	= 19 · 365[d]
6939	19[y] = 235[m] = 6939[d] 16;32,28,5[h] (Borg 1 3[b], 1—15)
7000	Second Coming of Christ
7007	= 143 · 49 = 13 · 532 + 91
7448	= 14 · 532
76	= 4 · 19
776	776 + 8 · 778 = 7000
778	"time" (zaman), cf. 776
81517	number of words in 510 Psalms
84	7000 = 13 · 532 + 84
875	875 · 8 = 7000
91	= 7 · 13, Enoch season
95	= 5 · 19, "great epact cycle"
958	= 13 · 76 = 52 · 19

2. *Sexagesimal*

1,41,45,29,8	1;41,45,29,8[d] = 28 · 0;3,38,3,11[d]
3,26,46,29,0,16,48	= 19 · 10;52,58,22,7,12 = 19 · *e*
3,38,3,11	19 · 6,5;15 — 3,55 · 29;31,50,8,9,24 = 0;3,38,3,11[d]
3,55	cf. 3,38,3,11
5,54,22	Arabic lunar year = 5,54;22[d]
5,54;22,1,35,30	cf. 29;31,50,7,57,30
5,54;22,1,37,52,40	emend to . . .,52,48 (EMML 215 8[b] II, 24)
5,54;22,1,37,52,48	lunar year (EMML 215 4[a] II, 6/7; Vindob 6 3[a] 2/3)
6,5;14,32,30	tropical year (Ibn Yūnis; Borg 1 3[b] 1—15)
6,5;15,11	sidereal year (EMML 215 4[a] II, 6/7)
7;9,6,20	cf. 194311
8;48,38,12	cf. 29;31,50,7,57,30
10;52,58,22,7,12	epact (Vindob 6 3[a], 1—9)
22,9	cf. 194308
29;31,50,7,57,30	= 5,54;22,1,35,30/12
29;31,50,8,9,24	cf. 3,38,3,11 and 59,3,40,16,18,48
29;31,50,8,9,33,20	12 · 29;31,. . . = 5,54;22,1,37,52,48 (Marḥa 'E 367 I, 17—19)
59,3,40,16,18,48	= 29;31,50,8,9,24/30

VII. GLOSSARY

See also Abbreviations of Technical Terms (p. 12)

m, n means page m, note n

'āmata 'agār 125
  'ālam 116
  meḥrat 116
  samā'etāt 116
  warḫ 126
'anqaṣ 127
'aqmar 31,8
'azēb 200

bāḥra ḥasāb 166
baradyos, baryodes 165
barakat 162

čaraqā 126,47

daqīqa 177,11

'eda 175,21
'elata ṭabibān 78
'ēretrā 143,41

gedef 20,6; 162
gēnā 151,94
gizē 168

ḫasāb 68; 115
ḫasāba 'abaqtē 68; 71
  samāy wa meder 68; 222
ḥeḍuḍ 79,1

kefel 168
kēkros, kentros 175
leba meder 143
ledata 'abaqtē 191
libā 200

māḫtam 43
manbara 127
manfaq 139,24; 186
manfaqa manfaqu 36,28
mase'e 200
maṣḥafa milād 96,4
maṭqe'e **187**

sanbat 32; 127
śarqa lēlit/ma'ālt 139
śarqatito 35; 47; 219; 222
śegāhu 116
ṣere'e 35,24
subā'e 32

ta'amra 20
tarafa qamar 47
ṭārik 32
ṭentyon **218**
tito 35,26; 222,25

warḫ ḥeḍuḍa 79,1

zaman 174

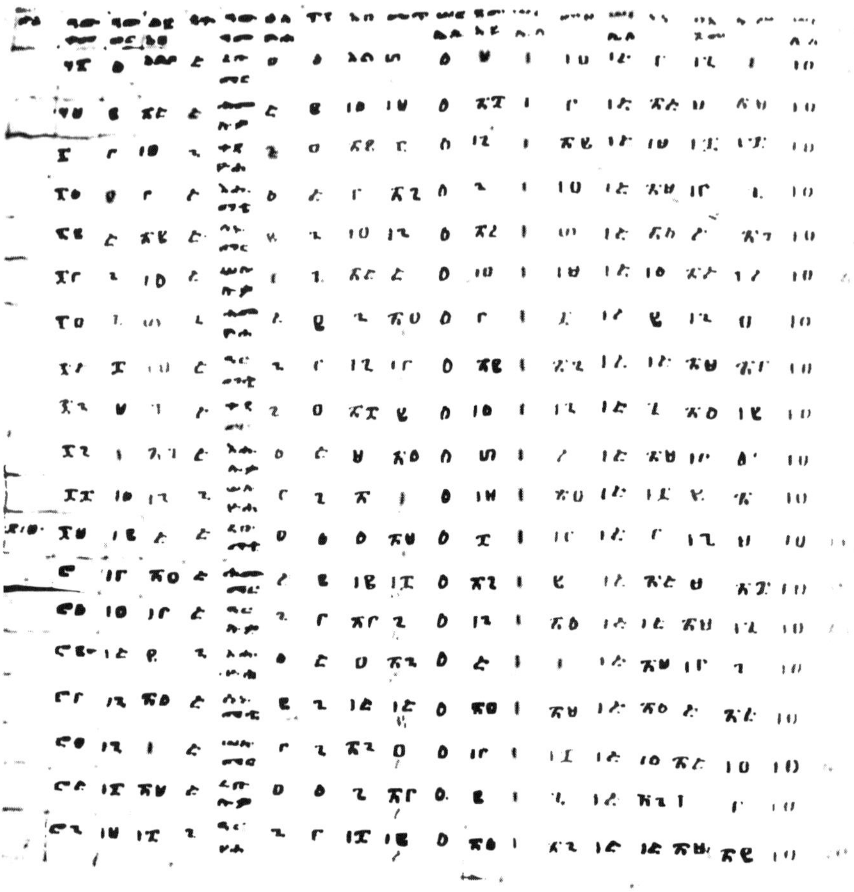

Ṭânâ 34 4b
k = 58 to 76

Plate I

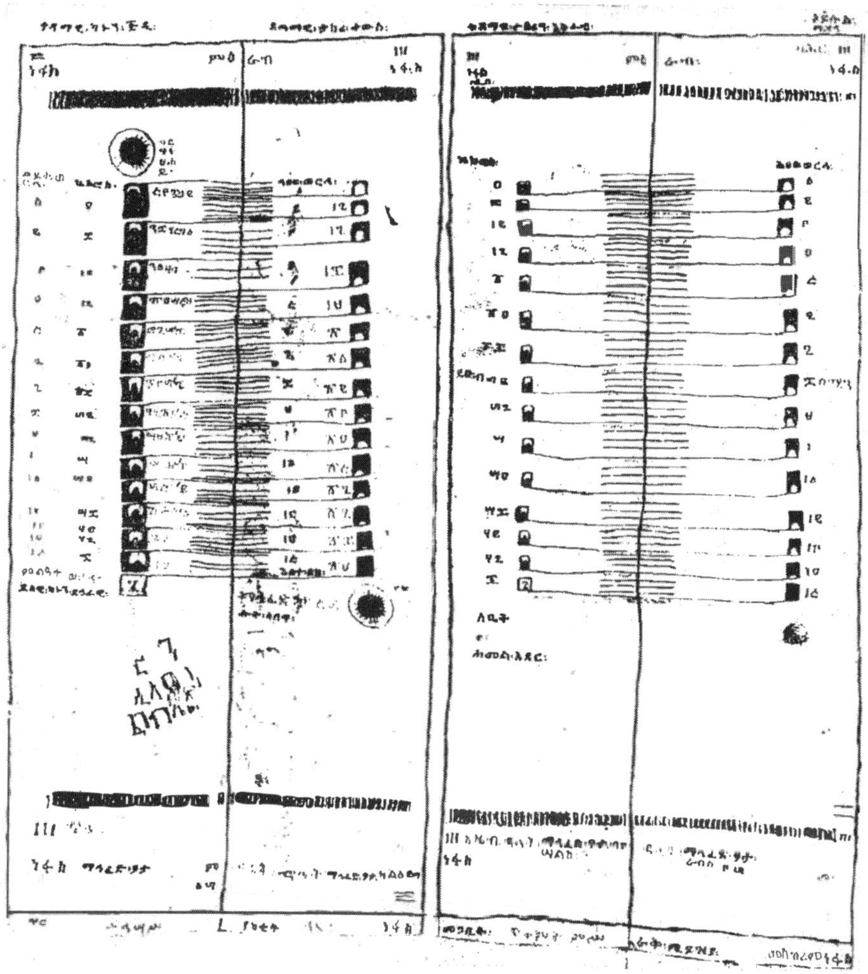

BM 809 3a

Plate II

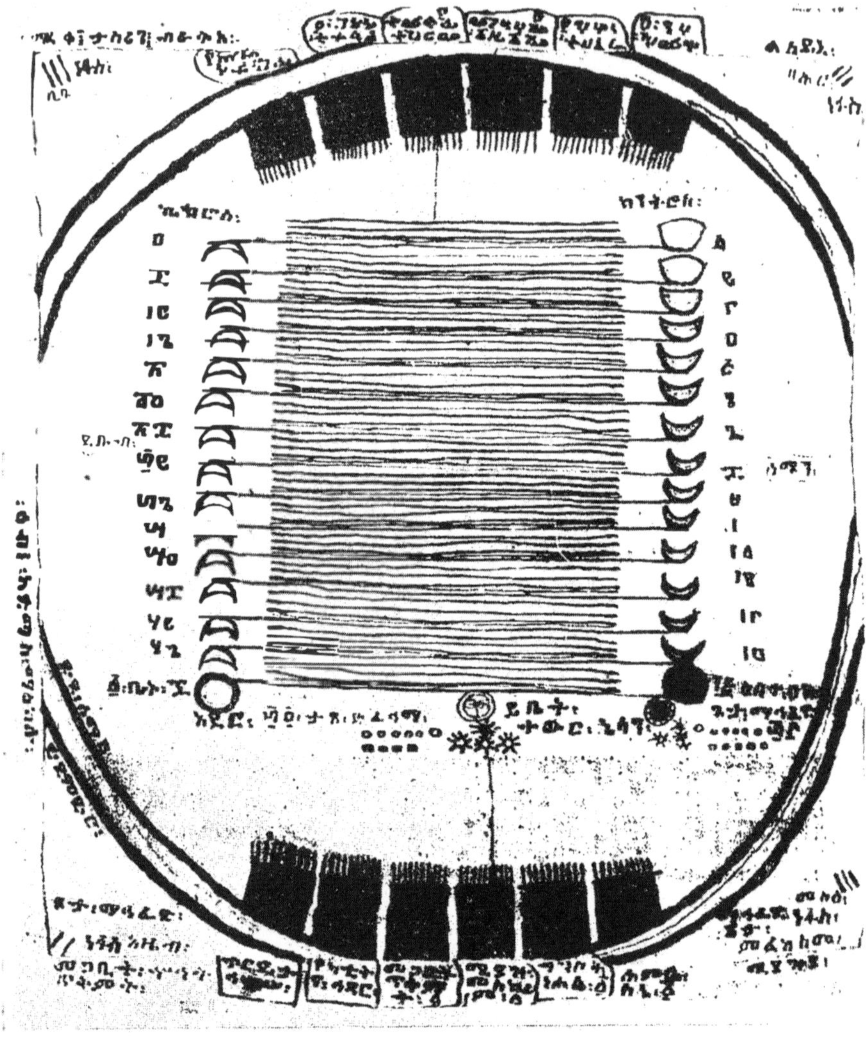

BM 816 16b

Plate III

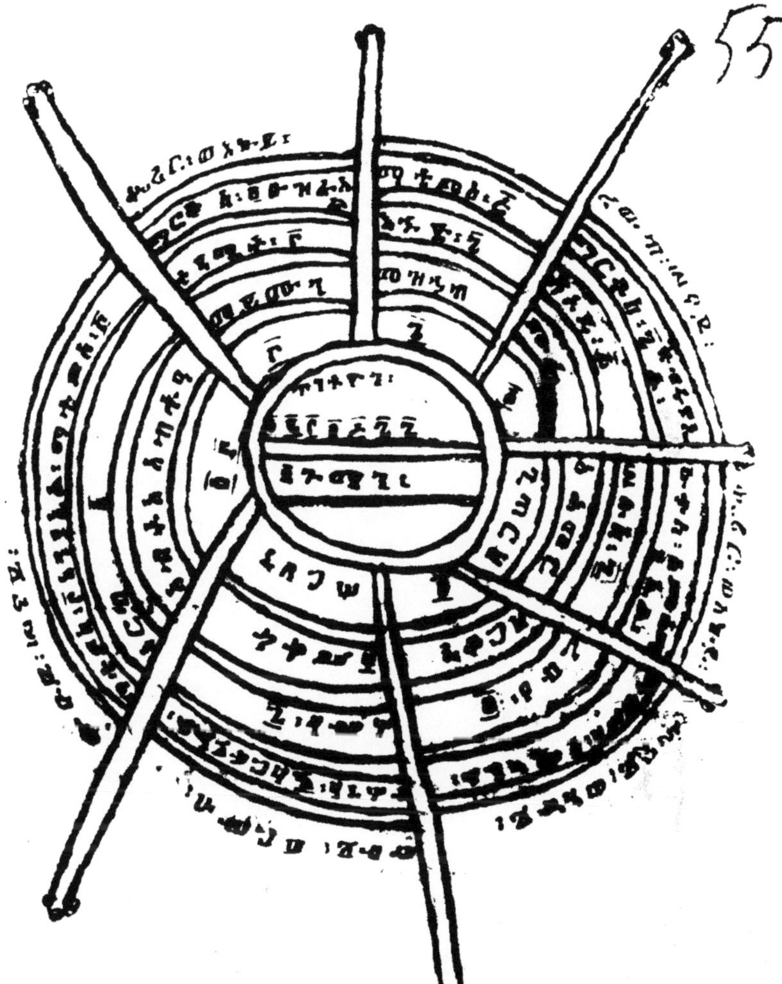

Vat 119 55a

Plate IV